T0134860

Studies in Systems, Decision and Control

Volume 110

Series editor

Janusz Kacprzyk, Polish Academy of Sciences, Warsaw, Poland
e-mail: kacprzyk@ibspan.waw.pl

About this Series

The series "Studies in Systems, Decision and Control" (SSDC) covers both new developments and advances, as well as the state of the art, in the various areas of broadly perceived systems, decision making and control- quickly, up to date and with a high quality. The intent is to cover the theory, applications, and perspectives on the state of the art and future developments relevant to systems, decision making, control, complex processes and related areas, as embedded in the fields of engineering, computer science, physics, economics, social and life sciences, as well as the paradigms and methodologies behind them. The series contains monographs, textbooks, lecture notes and edited volumes in systems, decision making and control spanning the areas of Cyber-Physical Systems, Autonomous Systems, Sensor Networks, Control Systems, Energy Systems, Automotive Systems, Biological Systems, Vehicular Networking and Connected Vehicles, Aerospace Systems, Automation, Manufacturing, Smart Grids, Nonlinear Systems, Power Systems, Robotics, Social Systems, Economic Systems and other. Of particular value to both the contributors and the readership are the short publication timeframe and the world-wide distribution and exposure which enable both a wide and rapid dissemination of research output.

More information about this series at http://www.springer.com/series/13304

Josep Domingo-Ferrer · David Sánchez
Editors

Co-utility

Theory and Applications

Editors
Josep Domingo-Ferrer
UNESCO Chair in Data Privacy, Department of Computer Science and Mathematics
Universitat Rovira i Virgili
Tarragona, Catalonia
Spain

David Sánchez
UNESCO Chair in Data Privacy, Department of Computer Science and Mathematics
Universitat Rovira i Virgili
Tarragona, Catalonia
Spain

ISSN 2198-4182 ISSN 2198-4190 (electronic)
Studies in Systems, Decision and Control
ISBN 978-3-319-86813-4 ISBN 978-3-319-60234-9 (eBook)
DOI 10.1007/978-3-319-60234-9

Printed on acid-free paper

This Springer imprint is published by Springer Nature
The registered company is Springer International Publishing AG
The registered company address is: Gewerbestrasse 11, 6330 Cham, Switzerland

Preface

How to guarantee that a global society without a common legal framework will operate smoothly? If charity and generosity do not arise spontaneously, one might design transactions so that helping others remains the best rational option.

This is precisely the goal of *co-utility*, which can be defined in game-theoretic terms as any interaction between peers in which the best option for a player to maximize her utility is to make sure the other players also get a fair share of utility. Example utilities in the information society transactions are functionality, security, and privacy. A protocol or mechanism designed using the co-utility principle ensures that helping others is the best rational option even if players are selfish.

This contributed book has been written under the umbrella of project *Co-utility: Conciliating individual freedom and common good in the information society*, which was supported by the *Templeton World Charity Foundation*. The first chapters lay the theoretic foundations of co-utility, either when it naturally arises in peer-to-peer interactions or when artificial incentives should be added to compensate the influence of negative utilities. Co-utility is also discussed in the context of the literature on social choice. Then, several applications of co-utility within the information society are presented: anonymous keyword search, digital oblivion, P2P content distribution, ridesharing, and collaborative data anonymization.

However, the applicability of co-utility is not restricted to the above domains investigated within the *Co-utility* project. This book also includes contributions by some of the participants in the *1st Co-utility Workshop*, which took place in Tarragona on March 10 and 11, 2016. In particular, there are chapters exploring the application of co-utility to environmental agreements, business model design, and the collaborative economy.

Tarragona, Catalonia
March 2017

Josep Domingo-Ferrer
David Sánchez

Contents

Contributors

Salvador Barberà MOVE, Universitat Autònoma de Barcelona, and Barcelona GSE, Bellaterra, Catalonia, Spain

Dolors Berga Departament d'Economia, Universitat de Girona, Girona, Catalonia, Spain

Riccardo Bonazzi Institute of Entrepreneurship and Management, University of Applied Sciences Western Switzerland, Sierre, Switzerland

Josep Domingo-Ferrer UNESCO Chair in Data Privacy, Department of Computer Science and Mathematics, Universitat Rovira i Virgili, Tarragona, Catalonia, Spain

Enrique Estellés-Arolas Catholic University of Valencia San Vicente Mártir, Valencia, Spain

Oriol Farràs UNESCO Chair in Data Privacy, Department of Computer Science and Mathematics, Universitat Rovira i Virgili, Tarragona, Catalonia, Spain

Vincent Grèzes Institute of Entrepreneurship and Management, University of Applied Sciences Western Switzerland, Sierre, Switzerland

Sergio Martínez UNESCO Chair in Data Privacy, Department of Computer Science and Mathematics, Universitat Rovira i Virgili, Tarragona, Catalonia, Spain

David Megías Internet Interdisciplinary Institute (IN3), Universitat Oberta de Catalunya, Castelldefels, Catalonia, Spain

Bernardo Moreno Facultad de Ciencias Económicas y Empresariales, Departamento de Teoría e Historia Económica, Universidad de Málaga, Málaga, Spain

Dritan Osmani Department of Computer Engineering and Mathematics, Universitat Rovira i Virgili, Tarragona, Catalonia, Spain

David Sánchez UNESCO Chair in Data Privacy, Department of Computer Science and Mathematics, Universitat Rovira i Virgili, Tarragona, Catalonia, Spain

Jordi Soria-Comas UNESCO Chair in Data Privacy, Department of Computer Science and Mathematics, Universitat Rovira i Virgili, Tarragona, Catalonia, Spain

Abeba Nigussie Turi UNESCO Chair in Data Privacy, Department of Computer Science and Mathematics, Universitat Rovira i Virgili, Tarragona, Catalonia, Spain

Part I
Theory of Co-utility

Co-utility: Designing Self-enforcing and Mutually Beneficial Protocols

Josep Domingo-Ferrer, David Sánchez and Jordi Soria-Comas

Abstract Protocols govern the interactions between agents, both in the information society and in the society at large. Protocols based on mutually beneficial cooperation are especially interesting because they improve the societal welfare and no central authority is needed to enforce them (which eliminates a single point of failure and possible bottlenecks). In order to guide the design of such protocols, we introduce co-utility as a framework for cooperation between rational agents such that the best strategy for each agent is to help another agent achieve her best outcome. Specifically, in this chapter we study and characterize self-enforcing protocols in game-theoretic terms. Then, we use this characterization to develop the concept of co-utile protocol and study under which circumstances co-utility arises.

1 Introduction

A protocol specifies a precise set of rules that govern the interaction between agents performing a certain task; that is, it details the expected behavior of each agent involved in the interaction for the task to be successfully completed. For example, for vehicles to avoid collisions at an intersection, they must wait for the green light. Also, in information technology there are plenty of protocols: the Internet Protocol (IP) defines how to route information packets in the Internet, the MESI protocol [19] defines how to preserve coherence between cache memories in multiprocessor architectures sharing a main memory, etc.

For protocols to be effective, they must be adhered to. This is not problematic when the participating agent cannot deviate by design, such as in the MESI protocol, but it

J. Domingo-Ferrer (✉) · D. Sánchez · J. Soria-Comas
UNESCO Chair in Data Privacy, Department of Computer Science and Mathematics, Universitat Rovira i Virgili, Av. Països Catalans 26, 43007 Tarragona, Catalonia, Spain
e-mail: josep.domingo@urv.cat

D. Sánchez
e-mail: david.sanchez@urv.cat

J. Soria-Comas
e-mail: jordi.soria@urv.cat

J. Domingo-Ferrer and D. Sánchez (eds.), *Co-utility*, Studies in Systems, Decision and Control 110, DOI 10.1007/978-3-319-60234-9_1

becomes an issue when the agents are free to choose between following the protocol or not, as it happens with vehicles at a crossing regulated by traffic lights. Although free agents cannot be forced to follow a protocol, rational free agents can be persuaded to do so if the protocol is properly designed. Such properly designed protocols will be called self-enforcing in the sequel. Examples of self-enforcing protocols that can be found in the literature include those involved in rational multiparty computation [4], the shotgun clause [1] (which is a way for two rational agents to agree on the price of an item) and the Vickrey auction [29] (which is a kind of auction in which each rational agent truthfully reports her valuation), among others [23].

While self-enforcement is essential, we are interested in protocols offering more than that: we want self-enforcing protocols that result in mutual help between agents and we call them *co-utile protocols*. A prominent advantage of co-utile protocols is that they promote social welfare. To illustrate, consider an agent that is interested in querying a web search engine but does not want the search engine to learn her queries, because these may disclose her personal features or preferences. If there is another agent also interested in privacy-aware querying, both agents can exchange (some of) their queries (and results), thereby preventing the web search engine from accurately profiling either agent; this results in a mutually beneficial collaboration [6, 7].

Co-utile protocols can be crafted for scenarios where the interests of the agents are complementary (or can be made complementary by adding appropriate incentives), so that helping other agents becomes the best way of pursuing one's own interests. Similar ideas about adding artificial incentives to promote cooperation have been proposed whose scope is narrower and more ad hoc than the one of the co-utility framework. For example, in P2P networks for sharing of distributed resources (e.g., storage, computing, data, etc.), incentives are used to achieve self-enforcing collaboration and deter the so-called *free-riders* (that is, peers who use resources from others but who do not offer their own resources) [20]; incentives in this context take the form of better service [2], or some sort of virtual money [12] for those who contribute.

In this chapter, we first formalize the notions of protocol and self-enforcing protocol. Then we move on to define co-utile protocols. Since we are assuming rational agents who freely decide whether to adhere to the protocol or not, game-theoretic modeling arises as the most natural choice. The assumption of free rational agents is plausible in peer-to-peer (P2P) scenarios lacking a central authority and a common legal framework that can be used to enforce a specific behavior.

2 Self-enforcing Protocols

Since we focus on protocols, we first need to clarify what we understand by *protocol*. Loosely speaking, a protocol is a sequence of interactions among a community of agents, called steps, that are aimed at carrying out a certain task.

A formalization of the concept of protocol that is often used in computer science is based on *finite state machines*. A finite state machine is a mathematical model of computation. It consists of a set of states, one of which is the current state, and a set of transitions between states that are triggered by specific events and modify the current state. While a finite state machine nicely models a protocol (each step changes from one state to another), it fails to capture the behavior of rational agents who choose their actions with the aim of maximizing their utility.

With rational agents in mind, game theory [14, 18] seems the right mathematical model. This theory models interactions between self-interested agents that act strategically. An agent is self-interested if she defines a preference relation over the set of possible outcomes of the protocol. On the other hand, an agent acts strategically when she takes into account her knowledge and expectations about the state of the world and about other agents to decide on her strategy (the way she plays the game). Game theory identifies subsets of outcomes (a.k.a. solution concepts) that agents would be most interested in achieving. In this work the focus will be on equilibrium solutions, i.e., outcomes which rational agents have no motivation to deviate from.

In our formalization, a game is used to model all the possible interactions among agents in the underlying scenario. In particular, the game includes also those interactions among agents that are not desired. Then, a protocol is regarded as a prescription of a specific behavior in the underlying scenario, that is, a sequence of desired interactions.

The type of interaction between agents is a key point in the outcome of a game. Game theory can model several interaction types, including:

- *Simultaneous and sequential moves.* Moves in a game are called simultaneous if each agent chooses her move independently (unaware) of the other agents' moves. On the other side, moves are called sequential if, at the time of choosing a move, previous moves made by other agents are known (at least to some extent).
- *Perfect and imperfect information.* A sequential game (one with sequential moves) is said to be a perfect-information game if the agent about to make her move has complete knowledge on the previous moves made by the other agents. If the agent's knowledge on previous moves is only partial, the game is said to be an imperfect-information game.
- *Complete and incomplete information.* If the previous category referred to knowledge on previous moves, this category refers to agents' knowledge on the underlying game. In games with complete information, the payoff of each agent at each final state is known by all agents. In games with incomplete information (a.k.a. Bayesian games), an agent is uncertain about the payoffs of the other agents.

The actual formalization of a game depends on the type of interaction one wants to model. We focus on sequential games with perfect information [13], because this is a quite common and basic type of interaction between agents.

The formal definition of a sequential game with perfect information is as follows:

Definition 1 (*Perfect-information game*) A perfect-information game (in extensive form) is a tuple $G = (N, A, H, Z, \chi, \rho, \sigma, u)$, where:

- N is a set of n agents;
- A is a set of actions;

- H is a set of non-terminal choice nodes;
- Z is a set of terminal nodes, disjoint from H;
- $\chi : H \rightarrow 2^A$ assigns to each choice node a set of possible actions;
- $\rho : H \rightarrow N$ assigns to each choice node a player $i \in N$ who chooses an action at that node.
- $\sigma : H \times A \rightarrow H \cup Z$ is an injective map that, given a pair formed by a choice node and an action, assigns to it a new choice node or a terminal node;
- $u = (u_1, ..., u_n)$, where $u_i : Z \rightarrow R$ is a real-valued utility function for agent i on the terminal nodes.

A perfect-information game can be represented in the so-called *extensive form* as a tree where:

- Each non-terminal choice node is labeled with the name of the agent making the move;
- Each terminal node is labeled with the utility that each agent obtains when reaching it;
- Edges going out from a node represent the actions available to the agent making the move.

Although the focus has been placed on sequential games with perfect information, nothing has been said so far about the completeness of the information. Assuming complete information seems too restrictive when trying to model the interactions between a set of potentially unrelated agents. Fortunately, the above tree representation can easily accommodate both complete and incomplete-information games:

- In games with complete information, the utilities at the terminal nodes are fixed values known to all the agents. Figure 1 shows sequential adaptations of two well-known games: the Prisoners' Dilemma and the Battle of the Sexes [15]. Both are perfect-information and complete-information games.
- In games with incomplete information, the utilities at terminal nodes are not completely known. This is modeled by replacing the fixed utilities at terminal nodes by utility functions that depend on an additional parameter: the type of each agent. The type of an agent encapsulates all the information on that agent that is not common knowledge. The set of types of the game is the Cartesian product of the set of types of each agent: $\Sigma = \Sigma_1 \times \ldots \times \Sigma_n$. Each agent knows her type but is uncertain about the types of the other agents. The agent models this uncertainty by attributing to every other agent a prior distribution over the possible types. Usually the same prior distribution is assumed for all agents.

We have defined a game as containing the set of actions, A, that are available to the agents. The strategies of a game describe the possible ways in which each agent can choose actions. The most basic type of strategies are pure strategies, defined as follows.

Definition 2 (*Pure strategy*) Let $G = (N, A, H, Z, \chi, \rho, \sigma, u)$ be a perfect-information game in extensive form. A pure strategy for an agent i is a function

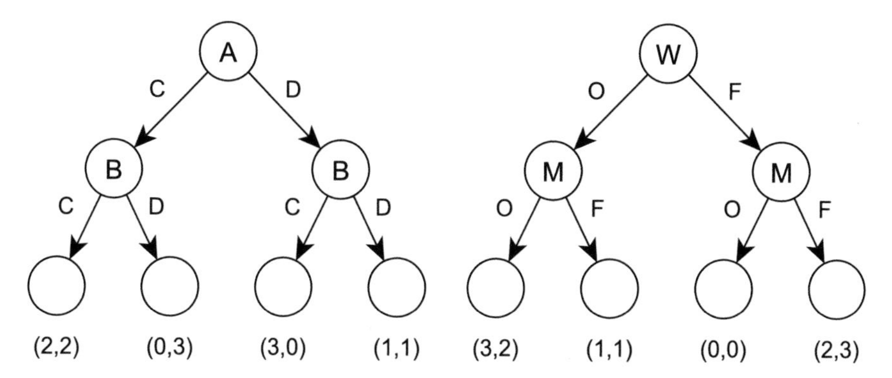

Fig. 1 Sequential versions of two well-known simultaneous games: the Prisoners' Dilemma (*left*) and the Battle of the Sexes (*right*). In the first game, prisoner A first decides whether to cooperate (C) with prisoner B, by *not* betraying him, or defect (D), by betraying him; then prisoner B faces the same decision (unawares of A's move in the simultaneous version). The utility of each prisoner is the reduction of the number of years in prison she or he obtains. In the second game, the woman, W, in a couple chooses between going to the opera (O) or to a football match (F); then the man, M, does the same (unawares of W's choice in the simultaneous version). The utility is how much they enjoy themselves: both would like to be together, but W prefers opera and M football

s_i that selects an available action at each non-terminal node at which i has to make a decision, that is, at each $h \in H$ such that $\rho(h) = i$. We denote by S_i the strategy set of agent i, that is, the set of pure strategies available to agent i.

In other words, a pure strategy provides a complete definition of how an agent will play a game, because it determines the move the agent will make for any situation he or she could face. Beyond pure strategies, more sophisticated strategies can be defined, such as mixed and correlated strategies, in which the selection of actions at each decision node h is randomized. We can now define the notion of strategy profile.

Definition 3 (*Strategy profile*) In a game G with n agents, a strategy profile s is a tuple $s = (s_1, \ldots, s_n)$, where s_i is the strategy chosen by agent i. The set of possible strategy profiles is $S_1 \times \ldots \times S_n$, where S_i is the set of strategies available to agent i.

As introduced above, after modeling the underlying scenario as a game, we can now model a protocol as a sequence of prescribed interactions. That is, a protocol embodies the agents' behavior that the designer wants to favor.

Definition 4 (*Protocol*) Given a perfect-information game G, a protocol P is a set of strategy profiles. If, at each non-terminal node h, agent $\rho(h)$ is allowed only one action by the protocol, then the protocol consists of a single strategy profile.

If G is represented in extensive form as a tree, a protocol can be viewed as a subtree from the root to several leaves. If the protocol consists of a single strategy profile, then it can be viewed as a path from the root to a leaf.

We are interested in self-enforcing protocols, that is, in prescribing behaviors that are adhered to by rational agents. In other words, no rational agent should have any incentive to deviate from the protocol provided that the other agents stick to it. This requirement can be rephrased in game-theoretic jargon by saying that each of the strategy profiles in the protocol must be an equilibrium of the underlying game. There is a large variety of equilibrium concepts such as Nash equilibrium, correlated equilibrium, subgame-perfect equilibrium, etc. In the context of protocols defined over perfect-information sequential games, we expect each agent to select a move that yields an equilibrium of the remaining subgame, that is, of the portion of the game that remains to be played after some moves have already been made by the agents. This kind of equilibrium is known as subgame-perfect equilibrium.

Definition 5 (*Subgame*) Given a game G in extensive form, the subgame of G rooted at node h is the restriction of G to h and its descendants.

Definition 6 (*Subgame-perfect equilibrium*) Given a game G in extensive form, a strategy profile s is a subgame-perfect equilibrium if, for any subgame G' of G, the restriction of s to G' is an equilibrium.

When computing Nash equilibria to check whether Definition 6 holds for them, in the case of complete-information games the given utilities are used, whereas in the case of incomplete-information games one resorts to expected utilities (derived from the prior distribution of types).

We are now in a position to formally define self-enforcing protocols using the concept of equilibrium:

Definition 7 (*Self-enforcing protocol*) A protocol P over a perfect-information game G is self-enforcing if no agent can increase her utility by deviating from one of the strategy profiles in P, provided that the other agents stick to that strategy profile. Equivalently, each of the strategy profiles in P is a subgame-perfect equilibrium of G.

In the sequential Battle of the Sexes (BoS) game described in Fig. 1, the protocol $P = (O, O)$ is self-enforcing: (O, O) is a Nash equilibrium of the BoS game, and (O) is a Nash equilibrium of the game that remains after the first action.

An interesting property of subgame-perfect equilibria is that they always exist; that is, every perfect-information game has a subgame-perfect equilibrium. Such an equilibrium is usually found by means of backward induction [14]. The backward induction algorithm assumes that, at each node, the agent making the move selects the action that gives her the best outcome. The algorithm starts evaluating the choice nodes that are parents of terminal nodes. Once the moves at these choice nodes have been selected, the algorithm proceeds backwards by evaluating the choice nodes that are parents of the previous choice nodes. For example, in the sequential BoS game (see Fig. 1), M is the agent making the last choice. In the left branch, the best option for M is O, which leads to utility (3, 2). In the right branch, the best option for M is F, which leads to utility (2, 3). Because W knows that M will seek to maximize his own utility, W can simplify the original tree to get a tree with a single choice node (hers) in which choosing O leads to utility $(3, 2)$ and choosing F leads to utility $(2, 3)$. Thus, in the original (unsimplified) tree with two decision stages, W should select O in the first place, and M should then also select O.

Let us find self-enforcing protocols in the two games proposed in Fig. 1. In the sequential version of the Prisoners' Dilemma, it is easy to see that the only self-enforcing protocol is (D, D). Choosing C is not self-enforcing for any agent because switching from C to D improves the agent's utility (provided that the other agent does not alter his/her move). In the sequential BoS game, there are two self-enforcing protocols: (O, O) and (F, F). In either protocol, no agent can improve her utility by deviating from the protocol provided that the other agent sticks to it. Of course, since in the sequential version agent W chooses her move first, she is most likely to favor (O, O) over (F, F).

3 Co-utility and Co-utile Protocols

As motivated in the introduction, we are interested in protocols that, beyond being self-enforcing, induce a rational agent to collaborate with other agents to increase their utilities. We say a protocol promotes collaboration between a set of agents if the utility of each participating agent is strictly greater than her utility when not participating (an agent is said to participate in the protocol if and only if she follows the prescriptions of the protocol). More precisely, we are interested in a protocol that promotes the most rewarding collaboration for all agents, that is, such that there is no alternative protocol whereby all agents could get a better outcome and at least one of them a strictly better outcome; in game-theoretic terms, we seek Pareto-optimality of the utilities of agents. This Pareto-optimal collaboration is the essence of co-utility, which can be defined as follows:

Definition 8 (*Co-utile protocol*) A self-enforcing protocol P is co-utile if each strategy profile in P is Pareto-optimal.

In the sequential BoS game described in Fig. 1, the protocol $P = (O, O)$ is co-utile. As we have already seen that it is self-enforcing, it only remains to check that it is Pareto optimal: W cannot improve her utility with respect to (O, O); M increases his utility if (F, F) is played, but that decreases the utility of W.

Co-utility is reminiscent of cooperative game theory [3] in that it deals with agents that collaborate to get a more desirable outcome. However, there is a substantial difference: in cooperative game theory, coalitions of agents are formed and the agents in a coalition coordinate their strategy to maximize the coalition's payoff, which is then divided among the agents. Assuming that the payoff can be divided among the agents in a coalition is essential for the coalitions to be formed. In contrast, in co-utility we assume that each agent acts autonomously and keeps to herself the payoff she gets: this allows co-utility to deal with non-divisible payoffs, such as privacy or security.

On the other hand, since the goal of co-utility is to lead to (co-utile) protocols, mechanism design is a natural framework to compare with. In mechanism design [16, 17], the ultimate goal is to come up with mechanisms that yield a previously defined socially desirable outcome. In this sense, co-utility is less demanding than mechanism design: it does not aim at enforcing a specific socially beneficial outcome, but rather at promoting a mutually beneficial collaboration between agents. It may well happen that the mutually beneficial outcome of a co-utile protocol is also socially desirable (but this does not necessarily happen).

Another difference between mechanism design and co-utility is that the former requires preference alignment, whereas the latter does not:

- In mechanism design, the outcome is selected based on the preferences reported by the agents. Untruthful reporting of preferences by agents in an attempt to obtain a more desirable outcome is an important issue that must be addressed by mechanism design. This is usually tackled by requiring some payment from agents that is calculated to make misrepresentation of utilities unfruitful. This kind of mechanism to align the preferences of the individual agents with the socially desirable outcome is known as incentive-compatible mechanism.
- In co-utility, rather than aligning preferences to the socially desirable outcome via incentives, we seek to promote collaboration between agents that have complementary preferences. For instance, for the case of privacy-preserving query submission to a web search engine, we assume that the agents that participate have complementary preferences regarding the topics they wish to query about, but they are all concerned about their privacy. Moreover, decisions in co-utility are made by the agents themselves rather than in a centralized way: when asked for collaboration, each agent decides which strategy is best for her.

In the sequential version of the Prisoners' Dilemma (Fig. 1), the only self-enforcing protocol is (D, D) but it is not Pareto-optimal because (C, C) provides a strictly greater payoff to both agents. Therefore, in this game no co-utile protocol exists. In the sequential version of the Battle of the Sexes (Fig. 1), it is easy to check that both self-enforcing protocols (O, O) and (F, F) are Pareto-optimal solutions of the game. Thus, both protocols are co-utile.

A specially interesting case of co-utility happens when the protocol is not only Pareto-optimal, but maximizes the utility of all agents:

Definition 9 (*Strict co-utility*) A protocol P on a game G is strictly co-utile if the utility that every agent derives from following any strategy in P is strictly greater than the utility the agent would obtain by following any strategy not in P.

We next prove that strict co-utility as per Definition 9 implies co-utility as per Definition 8.

Proposition 1 *If a protocol P on a game G is strictly co-utile, then it is co-utile.*

Proof We must check that P is self-enforcing and Pareto-optimal. Since all strategies not in P yield less payoff than any strategy in P, we have that:

1. No agent can increase her payoff by deviating from P and, thus, P is self-enforcing;
2. No protocol P' different from P yields a better utility to all agents and a strictly better utility to at least one agent; hence, P is Pareto-optimal.

Co-utility is about making selfish behavior compatible with mutually beneficial collaboration. However, given a game G, in general there is no guarantee that the selfish behavior of the agents will lead to co-utility, let alone strict co-utility, even if there are co-utile (or strictly co-utile) protocols in the game. However, for some specific games, such a guarantee can be given:

Proposition 2 *In a perfect-information game G where all the agents maximize their utilities in exactly the same set of terminal nodes of the tree that represents the game in extensive form (and only in these terminal nodes), selfish behavior by the agents will cause them to follow a strictly co-utile protocol.*

Proof Let $\mathscr{M}$ be the set of terminal nodes where the utilities of all agents are maximized. Let l be the length of the shortest path from the root to a node in $\mathscr{M}$. We will prove that selfish behavior leads to a node in $\mathscr{M}$ by induction over l. If $l = 1$, there is a single action to be chosen and at least one node $m \in \mathscr{M}$ can be reached by the agent making the choice. This agent could certainly take a longer path (with more than one action to be chosen), but this path should equally lead to a node in $\mathscr{M}$ (otherwise the agent would obtain a suboptimal payoff). We assume that the proposition is satisfied when $l \leq k$ and we need to show that it is also satisfied for $l = k + 1$. To apply the induction hypothesis, we split the shortest path to $\mathscr{M}$ into two parts: P_1 (containing the k initial steps) and P_2 (containing the final step). The same reasoning used for $l = 1$ shows that the agent making the decision at $leading(P_2)$, the leading node of P_2, chooses to follow P_2. As we have determined the path that will be followed if $leading(P_2)$ is reached, we can simplify the game tree by removing all the subtrees rooted at $leading(P_2)$ and copying the utilities of m (which are the same as the utilities of any node in $\mathscr{M}$) to $leading(P_2)$. To show that m is reached in the original tree, it is enough to show that $leading(P_2)$ is reached in the simplified tree. But the latter is immediate by applying the induction hypothesis to P_1, because the terminal node of P_1 is $leading(P_2)$.

A co-utile protocol always prescribes selfish behaviors (because it is self-enforcing) but not all selfish behaviors need to be co-utile. The interest of Proposition 2 is that it describes a class of games in which selfish behavior by everyone always yields strict co-utility. This is indeed very helpful to find a co-utile protocol. The alternative to find co-utile protocols is to use the backward induction algorithm (see [14]) on the game tree in order to seek subgame-perfect equilibria. However, this algorithm involves substantial computational cost, because it needs to traverse all the nodes of the tree (from leaves to root).

4 Application Fields

Although presenting detailed applications of co-utility will be done in other chapters, we do list here some areas in which co-utility seems to be a useful notion. In some of these applications areas, the interests of agents are naturally complementary (and hence mutual collaboration is a natural way to proceed); in other areas, the agents' interests can be made complementary by properly designing their interactions and/or adding appropriate incentives.

- *Anonymous keyword search.* Web search engines have become an essential tool to find online resources. However, their use can reveal very detailed information about the interests of their users. In [8, 10, 11] a co-utile protocol for exchanging queries between users is proposed. By exchanging queries with other users, the query profile associated to each user is masked and approximates the average group interests. However, the query exchange protocol has a cost. To prevent free-riders (who take advantage of the protocol to mask their own profile but are reluctant to help others) [5] proposes to use a reputation management system (see also the next chapter in this book).
- *Collaborative anonymization.* In surveys collecting individual data (microdata), each respondent is usually required to report values for a set of attributes. If some of these attributes contain sensitive information, the respondent must trust the collector not to make an inappropriate use of the data. If the respondent does not trust the data collector (which is understandable, given the large number of data collectors in our big data society [25]), she can anonymize her data locally by reporting inaccurately or by reporting nothing at all. The sensitivity of a piece of data depends on its rareness within a population [21]. In this respect, local anonymization is clueless in determining the right amount of masking. In [24] a co-utile protocol is proposed whereby users interact to determine the level of sensitivity of their respective pieces of data.
- *Content distribution.* Multicast distribution of content is not suited to content-based electronic commerce because all buyers obtain exactly the same copy of the content, in such a way that unlawful redistributors cannot be traced. Unicast distribution has the shortcoming of requiring one connection with each buyer, but it allows the merchant to embed a different serial number in the copy obtained by each buyer, which enables redistributor tracing. Peer-to-peer (P2P) distribution is a third option that can combine some of the advantages of multicast and unicast: on the one hand, the merchant only needs unicast connections with a few seed buyers, who take over the task of further spreading the content; on the other hand, if a proper fingerprinting mechanism is used, unlawful redistributors of the P2P distributed content can still be traced. In [9], a co-utile fingerprinting mechanism for P2P content distribution if proposed; it allows redistributor tracing, while preserving the privacy of most honest buyers and offering collusion resistance and buyer frameproofness.
- *Distributed computing.* In distributed scenarios, such as cloud provider federations, the Internet of Things, etc., a number of autonomous or entirely independent

systems are supposed to interact in certain prescribed ways. Since those systems often operate in different legal jurisdictions, compelling them to perform as agreed (even if stated in a contract) may be difficult. Making prescribed interactions rationally sustainable would be hence interesting. Co-utile protocols induced by artificial incentives such as reputation [5] (see also the next chapter in this book) or quality of service could go a long way in this direction.

- *Collaborative economy*. Collaborative markets are booming, such as crowdsourcing [26, 27], P2P lending [28], ridesharing [22], etc. The lack of a common legal framework and, hence, the lack of trust among peers hamper a more generalized take-off of the collaborative economy. Especially interesting would be a fully distributed collaborative economy, to empower the individual agents and overcome the current oxymoron of some platform-owning companies acting as oligopolies in the collaborative market. Just as in the case of distributed computing, suitable artificial incentives like distributed reputation [5] (see also the next chapter) could be leveraged to advance towards a truly distributed collaborative economy.

5 Conclusions

We have presented and formalized *co-utility*, which is a new paradigm to design self-enforcing protocols that make mutual help between rational agents the best strategy. Co-utility has the potential to boost social welfare in a great number of interactions, both in information society and in the society at large.

We have presented a list of possible applications of co-utility. In all of them, the interests of the agents are either complementary or can be made complementary by crafting a specific protocol with appropriate incentives (e.g. reputation management).

Acknowledgements Funding by the Templeton World Charity Foundation (grant TWCF0095/AB60 "CO-UTILITY") is gratefully acknowledged. Also, partial support to this work has been received from the Government of Catalonia (ICREA Acadèmia Prize to J. Domingo-Ferrer and grant 2014 SGR 537), the Spanish Government (projects TIN2014-57364-C2-1-R "SmartGlacis", TIN2015-70054-REDC and TIN2016-80250-R "Sec-MCloud") and the European Commission (projects H2020-644024 "CLARUS" and H2020-700540 "CANVAS"). The authors are with the UNESCO Chair in Data Privacy, but the views in this work are the authors' own and are not necessarily shared by UNESCO or any of the funding bodies.

References

1. Brooks, R.R.W., Landeo, C.M., Spier, K.E.: Trigger happy or gun shy? Dissolving common-value partnerships with Texas shootouts. RAND J. Econ. **41**(4), 649–673 (2010)
2. Buragohain, C., Agrawal, D., Suri, S.: A game theoretic framework for incentives in P2P systems. In: Shahmehri, N., Graham, R.L., Caronni, G. (eds.) Peer-to-Peer Computing, IEEE Computer Society, pp. 48–56 (2003)

3. Chalkiadakis, G., Elkind, E., Wooldridge, M.: Cooperative game theory: basic concepts and computational challenges. IEEE Intell. Syst. **27**(3), 86–90 (2012)
4. Dodis, Y., Rabin, T.: Cryptography and game theory. In: Nisan, N., et al. (eds.) Algorithmic Game Theory, pp. 181–205. Cambridge University Press, Cambridge (2007)
5. Domingo-Ferrer, J., Farràs, O., Martínez, S., Sánchez, D., Soria-Comas, J.: Self-enforcing protocols via co-utile reputation management. Inf. Sci. **367**(C):159–175 (2016)
6. Domingo-Ferrer, J., González-Nicolás, Ú.: Rational behavior in peer-to-peer profile obfuscation for anonymous keyword search. Inf. Sci. **185**, 191–204 (2012)
7. Domingo-Ferrer, J., González-Nicolás, Ú.: Rational behavior in peer-to-peer profile obfuscation for anonymous keyword search: the multi-hop scenario. Inf. Sci. **200**, 123–134 (2012)
8. Domingo-Ferrer, J., Martínez, S., Sánchez, D., Soria-Comas, J.: Co-utility: self-enforcing protocols for the mutual benefit of participants. Eng. Appl. Artif. Intell. **59**, 148–158 (2017)
9. Domingo-Ferrer, J., Megías, D.: Co-utility for digital content protection and digital forgetting. In: Proceedings of the 15th Annual Mediterranean Ad Hoc Networking Workshop, MedHocNet 2016, pp. 1–7. IEEE, New York (2016)
10. Domingo-Ferrer, J., Sánchez, D., Soria-Comas, J.: Co-utility: self-enforcing collaborative protocols with mutual help. Prog. Artif. Intell. **5**(2), 105–110 (2016)
11. Domingo-Ferrer, J., Soria-Comas, J., Ciobotaru, O.: Co-utility: self-enforcing protocols without coordination mechanisms. In: Proceedings of the 2015 International Conference on Industrial Engineering and Operations Management, IEOM 2015, pp. 1–17. IEEE, New York (2016)
12. Friedman, E.J., Halpern, J.Y., Kash, I.A.: Efficiency and nash equilibria in a scrip system for P2P networks. In: Eigenbaum, J., Chuang, J.C.-I., Pennock, D.M. (eds.) EC'06, pp. 140–149. ACM, Boston (2006)
13. Kuhn, H.W.: Extensive games and the problem of information. In: Kuhn, H.W., Tucker, A. (eds.) Classics in Game Theory, pp. 193–216. Princeton University Press, Princeton (1953)
14. Leyton-Brown, K., Shoham, Y.: Essentials of Game Theory: A Concise, Multidisciplinary Introduction . Morgan & Claypool, San Rafael (2008)
15. Luce, R.D., Raiffa, H.: Games and Decisions: An Introduction and Critical Survey. Wiley, New York (1957)
16. Maskin, E.: Nash equilibrium and welfare optimality. Rev. Econ. Stud. **66**(1), 23–28 (1999)
17. Nisan, N.: Algorithmic mechanism design. In: Handbook of Game Theory, pp. 477–516. Elsevier, Amsterdam (2014)
18. Osborne, M., Rubinstein, A.: A Course in Game Theory. MIT Press, Cambridge (1994)
19. Papamarcos, M.S., Patel, J.H.: A low-overhead coherence solution for multiprocessors with private cache memories. In: Agrawal, D.P. (ed.) ISCA, pp. 348–354. ACM, Boston (1984)
20. Rahman, R., Vinkó, T., Hales, D., Pouwelse, J.A., Sips, H.J.: Design space analysis for modeling incentives in distributed systems. In: Keshav, S., Liebeherr, J., Byers, J.W., Mogul, J.C. (eds.) SIGCOMM, pp. 182–193. ACM, Boston (2011)
21. Sánchez, D., Martínez, S., Domingo-Ferrer, J.: Comment on 'Unique in the shopping mall: on the reidentificability of credit card metadata'. Science **351**, 1274 (2016)
22. Sánchez, D., Martínez, S., Domingo-Ferrer, J.: Co-utile P2P ridesharing via decentralization and reputation management. Transp. Res. Part C **73**, 147–166 (2016)
23. Schneier, B.: The value of self-enforcing protocols. In: ThreatPost. https://threatpost.com/value-self-enforcing-protocols-081009/72980/ (2009). Accessed 17 Jan 2016
24. Soria-Comas, J., Domingo-Ferrer, J.: Co-utile collaborative anonymization of microdata. In: Torra, V., Narukawa, Y. (eds.) MDAI. LNCS 9321, pp. 192–206. Springer, Heidelberg (2015)
25. Soria-Comas, J., Domingo-Ferrer, J.: Big data privacy: challenges to privacy principles and models. Data Sci. Eng. **1**(1), 21–28 (2015)
26. Turi, A.N., Domingo-Ferrer, J., Sánchez, D., Osmani D.: Co-utility: conciliating individual freedom on common good in the crowd based business model. In: Proceedings of the 12th International Conference on e-Business Engineering, ICBE 2015, pp. 62–67. IEEE, New York (2015)
27. Turi, A.N., Domingo-Ferrer, J., Sánchez, D., Osmani, D.: A co-utility approach to the mesh economy: the crowd-based business model. Rev. Manag. Sci. (2016, in press)

28. Turi, A.N., Domingo-Ferrer, J., Sánchez, D.: Filtering P2P loans based on co-utile reputation. In: Proceedings of the 13th International Conference on Applied Computing, AC 2016, pp. 139–146 (2016)
29. Vickrey, W.: Counterspeculation, auctions, and competitive sealed tenders. J. Finance **16**(1), 8–37 (1961)

Incentive-Based Co-utility: Co-utile Reputation Management

Josep Domingo-Ferrer, Oriol Farràs and David Sánchez

Abstract Well-designed protocols should be self-enforcing, that is, be such that rational participating agents have no motivation to deviate from them. In addition, protocols can have other interesting properties, such as promoting collaboration between agents in a search for a better outcome. In [7, 8], we proposed the notion of *co-utility*, which characterizes a situation in which mutual help is the best rational option to take even for purely selfish agents; in particular, if a protocol is co-utile, it is self-enforcing. However, guaranteeing self-enforcement, let alone co-utility, for any type of agent behavior is not possible. To tackle this issue, in this chapter we detail how reputation mechanisms can be incorporated into existing protocols in order to make them self-enforcing (and optionally co-utile). Moreover, we show how to adapt and extend the well-known EigenTrust reputation calculation mechanism so that: (i) it can be applied to a variety of scenarios and heterogeneous reputation needs and, (ii) it is itself co-utile, and hence selfish agents are interested in following it. Obtaining a co-utile reputation mechanism creates a "virtuous circle" because: (i) the reputation management is self-enforcing and, (ii) as a result, it can be used to turn protocols that were not self-enforcing (resp. co-utile) per se into self-enforcing (resp. co-utile) ones.

1 Introduction

A protocol can be defined as a sequence of actions prescribed for an interaction between agents, both in the real world (e.g. road traffic rules) and the virtual world (e.g. peer-to-peer computing). A well-designed protocol is such that the

J. Domingo-Ferrer (✉) · O. Farràs · D. Sánchez
UNESCO Chair in Data Privacy, Department of Computer Science and Mathematics,
Universitat Rovira i Virgili, Av. Països Catalans 26, 43007 Tarragona, Catalonia, Spain
e-mail: josep.domingo@urv.cat

O. Farràs
e-mail: oriol.farras@urv.cat

D. Sánchez
e-mail: david.sanchez@urv.cat

J. Domingo-Ferrer and D. Sánchez (eds.), *Co-utility*, Studies in Systems,
Decision and Control 110, DOI 10.1007/978-3-319-60234-9_2

participating agents are willing to follow it, that is, they find no motivation to deviate from the protocol prescriptions. In this manner, the protocol can be completed in a self-enforcing way. Ensuring this in the general case (i.e., for any type of agent behavior) is not possible. This is why protocols are usually designed for the most usual type of agents: *rational agents*. A rational –self-interested– agent acts in order to maximize her profit/utility (i.e., the outcome she obtains, such as money, functionality, etc.). Thus, she will only deviate from a given protocol if by doing so she can increase her utility.

While being self-enforcing is essential for a protocol to be adhered to, other desiderata can be conceived. In particular, a protocol might promote mutually beneficial collaboration between agents, in the sense that following it entails a collaboration between agents that improves their utilities with respect to the non-collaborative scenario. This self-enforcing collaboration is captured by the notion of *co-utility* [7, 8] (see also the previous chapter in this book). We say that a protocol is *co-utile* if helping the other agents that participate in the protocol to increase their utilities is also the way to increase one's own utility. That is, the protocol is built on the notion of mutual help as the best rational option.

We can find situations in which interactions are naturally self-enforcing and even co-utile. However, in many real-life cases, "negative" incentives (e.g., costs, lack of privacy, fear of strangers, etc.) may override positive incentives for the agents to follow the rules, let alone help each other. To tackle this problem, artificial positive incentives may need to be added to the agents' utility functions in order to compensate negative incentives. Specifically, we detail a general distributed mechanism that provides reputation as an artificial positive incentive (e.g. a stranger may be less feared if she has a good reputation). It turns out, though, that managing the reputation of the agents in a distributed way constitutes in itself a protocol that requires collaboration (e.g. to calculate, update and disseminate reputations); hence, *it is crucial that reputation management be designed to be co-utile*, so that the collaboration it implies be rationally sustainable. This is a significant aspect that differentiates our reputation protocol from other reputation management solutions [12, 16], which rely on a central trusted authority [11], are application-oriented (e.g., file sharing, commercial transactions, social networks content generation) [1, 3, 25, 26] and/or just offer robustness against some tampering attacks [12, 13], but do not necessarily ensure the protocol to be rationally followed by the agents involved in the reputation calculation.

In this research, we first characterize the obstacles to designing self-enforcing and/or co-utile protocols that could be solved by appropriate reputation mechanisms; at the same time, we identify what a reputation mechanism should offer in order to qualify as appropriate. We then describe our general-purpose distributed reputation model inspired in the well-known EigenTrust [13] mechanism. However, unlike EigenTrust, which is meant to filter inauthentic content in P2P file sharing scenarios, *our reputation protocol can be applied to a variety of scenarios and heterogeneous reputation needs, and it is itself co-utile, so that self-interested agents wish to follow it*. As a result of the latter, we create a "virtuous circle": our co-utile reputation management is the result of self-enforcing, mutually beneficial collaboration and provides reputations that are used as artificial positive incentives to turn protocols that were not originally self-enforcing (resp. co-utile) into self-enforcing (resp. co-utile) ones.

2 Towards Co-utile Protocols via Reputation Management

Co-utility or even strict co-utility [7, 8] (see also the previous chapter in this book) can be reached in many practical situations. In the real world, for example, sharing one's car with one or several passengers can result in mutual benefits: passengers are able to reach their destination faster and the car driver needs to support less expense for the trip (because toll or gas fees are split among the passengers or because the driver can use facilities reserved for high occupancy vehicles or because of both reasons) [2]. In the information society, many peer-to-peer protocols follow the same pattern. File sharing protocols are based on the premise that, by sharing your own files and your upload bandwidth, you will also have access to a larger catalogue of files (shared by other users) with faster download speeds and greater availability than in centralized data storage systems [17]. Peer-to-peer protocols can also provide privacy protection to Internet users. For example, a user of a web search engine such as Google can hide her queries and search patterns from the search engine by requesting other peers to submit her queries (and return the results to her); this may be not only good for the requester's privacy, but also for the submitters' privacy, because the requester's query can be viewed as noise concealing the actual search interests of the submitters to the search engine [6, 9].

In [4, 7, 18, 21, 22] we showed how some of these protocols can be made co-utile. In all cases, this was possible because of the availability of agents with the appropriate types, that is, an appropriate combination of preferences/utilities (e.g., cost/time savings, privacy, functionality, etc.) that made their collaboration mutually beneficial. However, in practice, agent types can be very heterogeneous and "negative" utilities may dominate the benefits of agent collaboration, thereby precluding co-utility. For example, in a ride-sharing scenario, agents may be reluctant to share a car if the passengers and/or the driver are complete strangers [18]; in a query submission scenario, agents may be reluctant to submit queries from other users if that costs them bandwidth or money [5]; in a P2P lending or crowdfunding scenario, lenders and funders may be reluctant to given money to borrowers or projects with unknown reliability records [23, 24]. Another possible obstacle to co-utility occurs when the self-enforcing behavior is *not* to collaborate rather than collaborate. For example, in an uncontrolled file sharing system, purely selfish agents would prefer to download files from others but not to share their own files (to avoid spending upload bandwidth), thus becoming "free riders"; eventually, if all agents become free riders, the file sharing system collapses.

Obstacles like the above ones should be tackled if one wants to design protocols that are co-utile for as many agent types as possible. A way to neutralize the negative utilities that prevent co-utile collaboration is to incorporate artificial utilities in the form of rewards (that compensate the negative payoffs) or penalties (that discourage free riding). Note that rewards and penalties can also be used if, rather than obtaining a co-utile protocol, one merely wants to turn into self-enforcing a protocol that is not so.

Reputation is a very versatile artificial utility that can be used both as a reward and a penalty. In general, the reputation of an agent is the opinion of the community on the agent. On the one hand, reputation allows peers to build *trust*, which can neutralize the negative utilities related to mistrust (e.g., fear or reluctance in front of strangers). On the other hand, reputation also makes the agents accountable for their behavior: if an agent misbehaves, her reputation worsens and the other agents mistrust her more and more and are less and less interested to interact with her. In this manner, free riders and also malicious peers (who may try to subvert the system, even irrationally) may be identified and penalized (e.g., through limitation or denial of the service they try to abuse).

Many reputation mechanisms have been proposed in the literature for peer-to-peer communities, either centralized or distributed (see the surveys [12, 16]). However, since in our work reputation is used as a means to turn a protocol into co-utile or, at least, self-enforcing, the reputation calculation/management protocol itself should be distributed and strictly co-utile; otherwise, computing reputations would not be rationally sustainable and would not serve its purpose of inducing co-utility or self-enforcement in other decentralized P2P protocols.

Specifically, we want a reputation protocol with the following features, which should make it amenable to strict co-utility:

- *Decentralization*. Reputations should be computed and enforced by the peers themselves rather than a central authority. In this manner, we avoid depending on a trusted third party [11], which may be compromised and constitute a central point of failure (e.g., rational peers may be interested to attack this central authority).
- *Anonymity*. Reputation management should not rely on identifiers (e.g. IP addresses) that reveal the real identity of agents who help computing the reputation of other agents. Otherwise, the privacy loss may negatively affect the payoffs of collaborating in reputation management. Moreover, the possibility of creating coalitions between agents that know each other may facilitate collusion attacks to the reputation system. Surprisingly, most related works either support very limited anonymity or completely neglect it (see [19]).
- *Low overhead*. Reputation management should not imply a large expenditure of resources (e.g., bandwidth, storage, calculation); otherwise, these negative payoffs may dominate the benefits brought by reputation. To be more specific, a linear [13] or quasi-linear [20] calculation cost would be desirable.
- *Proper management of new agents*. Newcomers should not enjoy any reputation advantage; otherwise, malicious peers may be motivated to create new anonymous identifiers after abusing the system in order to regain the advantages of a good reputation.
- *Attack tolerance*. A number of attacks may be orchestrated in order to subvert the reputation system. Since we focus on rational selfish agents, we are interested in

systems that are robust against "rational attacks". In particular, a rational selfish agent may try to subvert the protocol or manipulate the reputation calculation if, by doing so, she can increase her benefit; that is, if via subversion she can obtain a higher reputation at a lower cost than via "good behavior" in the underlying P2P scenario. Thus, the reputation system should implement measures to make the cost of such an attack unattractively high. According to the classification of attacks identified in [12], we want to avoid the following attacks:

- *Self-promotion*, where agents are able to falsely increase their own reputations at a small or zero cost.
- *Whitewashing*, in which agents circumvent the consequences of abusing the system to obtain an unfair benefit, for example by creating new "clean" identities or performing Sybil attacks. Analyses of whitewashing and Sybil attacks can be found in [10, 15], respectively.
- *Slandering*, where agents may falsely lower the reputation of other agents if, by doing so, their own reputation becomes comparatively higher.
- *Denial of service*, in which agents block the calculation and dissemination of reputation values. This may happen, for example, if the reputation calculation has a cost that agents deem higher than the benefits the calculation of their own reputation (by other agents) would bring to them.

After examining a number of decentralized reputation mechanisms [12], in [5], we selected, adapted and extended EigenTrust [13] in order to obtain our strictly co-utile reputation protocol. EigenTrust offers most of the desirable features identified above: distributed reputation calculation, low overhead, anonymity and robustness to attacks. This reputation scheme is designed to filter out inauthentic content in peer-to-peer file sharing networks. Its basic idea is to calculate a global reputation for each agent based on aggregating the local opinions of the peers that have interacted with the agent. If we represent the local opinions by a matrix whose component (i, j) contains the opinion of agent i on agent j, the distributed calculation mechanism computes global reputation values that approximate the left principal eigenvector of this matrix. Anonymity is achieved by means of a distributed hash table (DHT) that randomly links agents and reputation calculation duties. For unstructured networks, other solutions can be used, such as gossip-based reputation distribution protocols [27].

3 Reputation Calculation Model

The aim of the reputation mechanism is to compute global reputation values in a distributed way. These values are calculated by means of a protocol whereby the agents share their local reputation values. In this section, we extend EigenTrust to compute reputations based on non-binary opinions. The original EigenTrust system was designed for P2P file sharing, and the receiver's opinion on a file download is just binary: either the download has been satisfactory or not. We want to be able to compute reputations based on opinions that have several categories, or are even continuous. The continuous case may also accommodate a situation in which each agent quantifies the benefit/cost resulting from interacting with another agent.

3.1 Calculation of Local Reputations

The opinion of $\mathscr{P}_i$ on another agent $\mathscr{P}_j$ with whom $\mathscr{P}_i$ has directly interacted is the reputation s_{ij} of $\mathscr{P}_j$ local to $\mathscr{P}_i$. We define this value as the aggregation of payoffs (either positive or negative) that $\mathscr{P}_i$ has obtained from the set of transactions (Y_{ij}) performed with $\mathscr{P}_j$:

$$s_{ij} = \sum_{y_{ij} \in Y_{ij}} payoff_i(y_{ij}).$$

Payoffs may depend on the particular scenario, but the above calculation is general enough to accommodate them. Specifically, we can cope with the following scenarios that cover all the cases described in [12]:

- *Binary payoffs*. These are just binary values, either positive (+1) or negative (−1), according to whether the transaction with $\mathscr{P}_j$ has fulfilled or not $\mathscr{P}_i$'s goals. Binary payoffs make sense in P2P file sharing networks (i.e., the desired file has been obtained or not) or commercial transactions (e.g., in eBay, buyers rate transactions either as positive or negative).
- *Discrete or continuous opinions*. Payoffs may also measure the opinion on a transaction as a discrete or continuous magnitude. In this case, a certain transaction may generate reciprocal reputation updates by the involved agents (i.e., $\mathscr{P}_i$ on $\mathscr{P}_j$ and $\mathscr{P}_j$ on $\mathscr{P}_i$). Most transactions with a subjective outcome (other than the objective outcome of fulfilling or not the task/goal) should be measured in this way. For example, in a car-sharing network, passengers and drivers can rate each other to measure how pleasant the trip was or whether there were any unnecessary delays or overcosts; all of these dimensions can influence the utility of the involved peers and, thus, affect their willingness to collaborate in the future. Commercial transactions can also be rated in this way (e.g., Amazon provides discrete positive ratings), even though the rating is one-sided (not reciprocal). In some cases, the opinion can be the aggregation of several transaction components (e.g., reviews, comments, etc.) [14].
- *Costs*. In the previous cases, reputation values are potentially unbounded (i.e., there is not a limited reputation budget to be distributed among the peers of the network) and independent between the agents. Alternatively, payoffs could reciprocally measure the cost incurred/caused by the agents when helping others/being helped in fulfilling a task/goal. For example, in a P2P file sharing network, if $\mathscr{P}_i$ provides a file to $\mathscr{P}_j$, the payoff of the former with respect to the latter would be a negative value measuring the bandwidth spent by $\mathscr{P}_i$ when uploading the file; likewise, the payoff of $\mathscr{P}_j$ (who has received the desired file) with respect to $\mathscr{P}_i$ would

be the inverse positive value. In the privacy-preserving query submission scenario discussed in Sect. 2 a similar criterion can be applied: if performing a query to the database has a cost (e.g. money) for the submitter, this should be reflected as a positive reputation payoff for the submitter and as a negative reputation payoff for the query originator. In these scenarios, reputation values can measure the difference between the cost incurred by an agent $\mathscr{P}_i$ when helping others and the cost incurred by others when helping $\mathscr{P}_i$. A reputation greater than 0 shows that the agent has helped others more than what others have helped her. Unlike in the previous cases in which reputation values were unbounded, here reputations are positively/negatively balanced across the members of the network; thus, a reputation value around 0 indicates a fair allocation of costs among peers.

It is important for local reputation values computed by the different peers to measure the payoff of similar transaction outcomes in a similar way. For objective outcomes (such as the fulfillment of a task or the incurred cost), this is not problematic; however, for subjective opinions the designer of the reputation system may provide some rules to control the ratings. For example, eBay recently implemented a "subjective" rating[1] which buyers can use to rate sellers from 5 stars down to 1 star regarding several aspects of the transaction (e.g., delivery speed of the goods, shipping charges, etc.); however, eBay associates each value to an objective criterion (e.g., a 5-star shipping charge means free shipping, whereas less than 3-star means a charge higher than the real shipping cost).

3.2 Computing Global Reputations

We review here how EigenTrust computes global reputations from local reputations; recalling this is needed for the reader to understand what we propose in Sects. 4 and 5. In order to properly aggregate the local reputation values computed by each peer, a normalized version c_{ij} is first calculated as follows:

$$c_{ij} = \frac{\max(s_{ij}, 0)}{\sum_j \max(s_{ij}, 0)}.$$

In this manner, the normalized local reputation values lie between 0 and 1 and their sum is 1. In other words, each agent has a reputation budget of only 1 that she has to split among her peers proportionally to her positive experiences (negative experiences are truncated to 0). This makes all agents equal contributors to the global reputation and avoids dominance by agents with a larger number of experiences. Moreover, this normalized calculation deters peers from colluding by assigning arbitrarily high values to each other. Finally, the fact that negative reputation values are truncated to 0 prevents selfish agents from assigning arbitrarily low values to good peers. We will later discuss how the protocol can also withstand tampering with reputation values.

[1] http://pages.ebay.com/help/feedback/detailed-seller-ratings.html.

A side effect of the truncation of negative values is that reputation values do not distinguish between agents with whom $\mathscr{P}_i$ had a bad experience (negative local reputation) from those with whom $\mathscr{P}_i$ has not interacted so far. Even though this may be seen as a drawback, it also can be viewed as a strength: newcomers (i.e., agents who have not yet interacted with anyone) do not have any reputation advantage, because their reputation is indistinguishable from the one of misbehaving agents. As a result, a selfish agent has no incentive to take a new virtual identity in order to "clean" her reputation after abusing the system (e.g., refusing to help others, in order to minimize her own costs). Likewise, newcomers will become instantly motivated to positively contribute to the system in order to earn the minimum reputation that other agents would require from them.

We can thus see that normalizing local reputation values biases the system towards positive reputation; that is, agents need a minimum *positive* reputation value in order to be trusted by peers, which requires them to help others/contribute to the system first on order to be able to get help later.

In this setting, local reputation values are disseminated and aggregated through the network peers following a transitive reputation algorithm, in order to obtain the global reputation value of each agent. This is the main idea of EigenTrust. Any agent $\mathscr{P}_i$ can compute $\hat{t}_{ik}^{(0)}$, an approximation of the reputation of a potentially unknown peer $\mathscr{P}_k$, by asking the peers with whom $\mathscr{P}_i$ has interacted ($\mathscr{P}_j$) for their local reputation w.r.t. $\mathscr{P}_k$, that is c_{jk}. Since $\mathscr{P}_i$ has already computed the local normalized reputation w.r.t. $\mathscr{P}_j$, that is c_{ij}, $\mathscr{P}_i$ can compute a local estimate of the reputation t_{ik} of $\mathscr{P}_k$ by using c_{ij} to weight $\mathscr{P}_j$'s local reputation; specifically, $\hat{t}_{ik}^{(0)} = \sum_j c_{ij} c_{jk}$. Thanks to the local normalization, $\hat{t}_{ik}^{(0)}$ takes values between 0 and 1. Observe that if we call $\mathbf{c}_i = (c_{i1}, \ldots, c_{in})^T$ and $\mathbf{C} = [c_{ij}]$, then $\hat{\mathbf{t}}_i^{(0)} = \mathbf{C}^T \mathbf{c}_i$, where $\hat{\mathbf{t}}_i^{(0)} = (\hat{t}_{i1}^{(0)}, \ldots, \hat{t}_{in}^{(0)})^T$. If every agent $\mathscr{P}_i$ computes $\hat{\mathbf{t}}_i^{(0)}$, in the next iteration $\mathscr{P}_i$ can compute $\hat{\mathbf{t}}_i^{(1)} = \mathbf{C}^T(\mathbf{C}^T \mathbf{c}_i)$. After m iterations, $\mathscr{P}_i$ will compute $\hat{\mathbf{t}}_i^{(m-1)} = (\mathbf{C}^T)^m \mathbf{c}_i$. Under the assumptions that $\mathbf{C}$ is irreducible and aperiodic [13], in an ideal and static setting the succession of reputation vectors computed by any peer will converge to the same vector for every peer, which we call $\mathbf{t} = (t_1, \ldots, t_n)^T$ and is the left principal eigenvector of $\mathbf{C}$. The j-th component of $\mathbf{t}$ represents the global reputation of the system on each agent $\mathscr{P}_j$.

Computing the global reputation values by the above method is not efficient because the communication complexity is very high. In the following section we present a protocol to compute global reputation values that emulates the above method in a distributed and more efficient way.

4 Co-utile Distributed Reputation Calculation Protocol

In [5], we adapt and extend the EigenTrust secure protocol [13] in order to obtain a co-utile distributed protocol for computing global reputation values. The core idea of the EigenTrust secure protocol is that the reputation value of an agent $\mathscr{P}_i$ is computed by other agents in order to prevent manipulation and minimize the action of malicious peers. This computation is based only on the local reputation reported by those agents with whom the agent $\mathscr{P}_i$ interacted directly.

In our proposal, the computation of t_i (the global reputation of agent P_i) is based on the experience of the agents in A_i, which is the set of agents that provided help to or received help from $\mathscr{P}_i$. Hence, the global reputation values are computed in a distributed way according to the experiences of the agents in the network.

We assume that each agent $\mathscr{P}_i$ has an initial reputation value $t_i^{(0)}$ that is based on previous experiences or is given by default. Each agent has a number M of score managers that will compute her reputation value. These score managers are defined according to a distributed hash table (DHT), which maps each agent to a set of several agents determined by hash functions $h_0, h_1, \ldots, h_{M-1}$. In this way, $h_0(ID_i), \ldots, h_{M-1}(ID_i)$ are the coordinates of the agents computing the reputation value of $\mathscr{P}_i$.

With the above hash mapping, on average every agent is the score manager of M agents, so the work of the agents in the reputation mechanism is balanced. Let D_i be the set of agents for whom $\mathscr{P}_i$ is a score manager; we will also call D_i the set of daughters of $\mathscr{P}_i$. During the computation of the reputation values, the score manager of $\mathscr{P}_d$, say $\mathscr{P}_i$, learns A_d, that is, the set of agents that provided help to or received help from $\mathscr{P}_d$. Then $\mathscr{P}_i$ receives the trust assessments of $d \in D_i$ sent by the agents in A_d. The terms c_{ji} for $j \notin A_i$ are zero. Then, $\mathscr{P}_i$ engages in an iterative refinement to compute the reputation of every $\mathscr{P}_d \in D_i$, based on the score c_{jd} on $\mathscr{P}_d$ by each $\mathscr{P}_j \in A_d$ weighted by the score managers of $\mathscr{P}_j$ (the weight is the current reputation $t_j^{(k)}$ of $\mathscr{P}_j$ held by the score managers of $\mathscr{P}_j$). As a result, no agent $\mathscr{P}_j$ can directly influence the reputation of any other $\mathscr{P}_d$; everything is mediated by the score managers of $\mathscr{P}_j$, who together act as a sort of distributed trusted third party.

The above computations are described in Protocol 1. The main differences between this protocol and the one in [13] are: (i) we rely for most of the calculation on the score managers (Step 7), thereby increasing the redundancy of the computation of the reputation values and protecting this computation against malicious agents; and (ii) our definition of the agents that provide the local reputation values is more general, since we include any agent that has interacted with the daughter agent (Steps 3, 5, 9, 10 and 11).

Once a reputation manager computes a reputation on a daughter agent using Protocol 1, she keeps the reputation value for that agent until the next protocol execution. Protocol 1 is meant to be run periodically, in order for reputations to stay up-to-date. The reputation update period can be set depending on the activity of the agents, in order to obtain faster updates when the frequency of agent interactions increases. Ideally, the protocol should be run in parallel and asynchronously with respect to the agent interactions.

After the computation of the global reputation values, if an agent $\mathscr{P}_i$ needs the reputation value t_j of $\mathscr{P}_j$ in order to decide whether to collaborate with him, she can query the M score managers of $\mathscr{P}_j$ for his reputation. The M values obtained from

Protocol 1 CO- UTILE COMPUTATION OF REPUTATION VALUES

1: **for all** agent $\mathscr{P}_i$ **do**
2: Submit local reputation values $\mathbf{c}_i$ to all score managers at positions $h_m(ID_i)$, for $m = 0, \ldots, M-1$;
3: Collect local reputation values $\mathbf{c}_d$ and the set A_d for all daughter agents $\mathscr{P}_d \in D_i$;
4: **for all** $\mathscr{P}_d \in D_i$ **do**
5: Query all the score managers of $\mathscr{P}_j \in A_d$ for $c_{jd}t_j^{(0)}$
6: $k := -1$
7: **repeat**
8: $k := k + 1$
9: Compute $t_d^{(k+1)} = c_{1d}t_1^{(k)} + c_{2d}t_2^{(k)} + \ldots + c_{nd}t_n^{(k)}$;
10: Send $c_{dj}t_d^{(k+1)}$ to all the score managers of $\mathscr{P}_j \in A_d$;
11: Wait for all score managers of $\mathscr{P}_j \in A_d$ to return $c_{jd}t_j^{(k+1)}$;
12: **until** $|t_d^{(k+1)} - t_d^{(k)}| < \varepsilon$ // Parameter $\varepsilon > 0$ is a small value
13: **end for**
14: **end for**

the M score managers should be the same, because the inputs of the score managers are the same. However, if some values are different (e.g., if some score managers or agents have altered the computation for some reason), $\mathscr{P}_i$ can take as t_j the most common value among the ones sent by the score managers. If a score manager $\mathscr{P}_k$ does not answer to a query by $\mathscr{P}_i$, then $\mathscr{P}_i$ will assume that $\mathscr{P}_k$ is not active and she will set $c_{ik} = 0$.

In addition to the global reputation, the experience of $\mathscr{P}_i$ with respect to $\mathscr{P}_j$ is also reflected in the local value c_{ij}, if available (i.e., if $\mathscr{P}_i$ and $\mathscr{P}_j$ have already interacted). In some cases, local and global reputation values may not be coherent because the latter is the aggregated version of the former. In general, it is better for $\mathscr{P}_i$ to consider both local reputations $\mathbf{c}_i$ and global reputations $\mathbf{t}$ to make decisions about collaboration; a conservative criterion would be to use the lowest among the local and global reputations.

4.1 Co-utility Analysis

Within the co-utility framework, we consider that agents are rationally selfish. For the calculation of reputations, this means that: (i) agents want their reputation to be computed (this is their main utility); (ii) they are interested in maximizing their reputation (so that it is higher than the reputations of other peers) by any means (i.e., either by correctly following the protocol or by deviating from it). Ideally, the reputation calculation protocol should be self-enforcing (hence discouraging deviation); what is more, it should be co-utile, given that all agents obtain their correct reputation as the outcome utility of the protocol execution. In the sequel, we analyze Protocol 1 and, for each step, we describe the options available to the agents and we justify why correctly following the protocol is the rational choice and, thus, why the protocol is self-enforcing and, in particular, co-utile.

Generally speaking, Protocol 1 encourages the agents to collaborate because the impact of their opinions on the computation of the global reputation values increases when they are active (that is, present in as many sets A_* as possible). On the other hand, the protocol is robust against *self-promotion attacks*, because the agents that actually compute the global reputation values of an agent $\mathcal{P}_i$ (her score managers) are chosen randomly, and they use the information provided by the score managers of the agents with whom $\mathcal{P}_i$ has interacted; thus, the agents cannot manipulate their own reputation values because they do not have direct control on their calculation. Also, Protocol 1 is robust against *whitewashing attacks*: since the local reputation values are truncated to 0, the system does not distinguish between agents with whom there was a bad experience (whose negative s_{ij} is truncated to 0) and those with whom no one has interacted so far. In this way, selfish agents have no incentive to create new virtual identities to reset their bad reputation because, in practice, this does not make them better off.

Let us now be more specific and go step by step. At Step 2, agent $\mathcal{P}_i$ acts as a reputation assessor and sends her local reputation values $\mathbf{c}_i$ to the reputation managers. Any peer $\mathcal{P}_i$ has a rational interest in forwarding to the rest of peers the fair local reputations c_{ij} she has awarded to peers $\mathcal{P}_j \in A_i$. We next justify why:

- If $\mathcal{P}_i$ fails to forward c_{ij}, then c_{ij} is taken as zero, and hence the reputation t_j of $\mathcal{P}_j$ is lower than due. Now, since $\mathcal{P}_j \in A_i$ implies $\mathcal{P}_i \in A_j$, by the expression in Step 9, a lower t_j also results in a lower reputation t_i of $\mathcal{P}_i$, *ceteris paribus*.
- If $\mathcal{P}_i$ forwards a reputation c'_{ij} less than the real c_{ij}, the argument is the same as in the previous item. Hence, Protocol 1 is robust against *slandering attacks by assessors*.
- If $\mathcal{P}_i$ forwards a reputation c'_{ij} greater than c_{ij}, this increases t_j and t_k for agents $\mathcal{P}_k \in A_j$. Although $\mathcal{P}_i$ is also in A_j and thus benefits from the increase, *ceteris paribus* t_i will increase less than t_j and similarly as t_k for $k \neq i, j$. Hence, $\mathcal{P}_i$ will be in a worse relative position with respect to $\mathcal{P}_j$ and possibly other peers.

From Step 3 onwards, agent $\mathcal{P}_i$ acts as a reputation manager and collects the local reputations awarded by all his daughter agents $\mathcal{P}_d$, which in turn allows $\mathcal{P}_i$ to learn the set A_d of peers with whom $\mathcal{P}_d$ has interacted. The same justification above that $\mathcal{P}_i$ is rationally interested to report fair local reputations ensures that $\mathcal{P}_d$ will report all her local reputations fairly. On the other hand, if the score manager $\mathcal{P}_i$ does not collaborate to compute the global reputation of his daughters (*denial-of-service attack*), the requester of t_d will receive no answer from $\mathcal{P}_i$; in this case, the requester will consider that $\mathcal{P}_i$ not active, and will remove $\mathcal{P}_i$ from the list of available agents. Hence, $\mathcal{P}_i$'s own reputation value will be decreased in the following iteration of the reputation calculation protocol. Thus, a rational $\mathcal{P}_i$ is not interested in denying service.

The iteration starting in Step 7 involves the score manager $\mathcal{P}_i$ and the score managers of peers in A_d. For the same reasons given above to justify that $\mathcal{P}_i$ is not interested in denying service, the score managers of peers in A_d are not interested either.

Finally, even if not denying service, a score manager might send a wrong value when queried by an agent (Steps 5,10 and 11). Assuming the security parameter M has been chosen so that the number of malicious agents can be safely considered to be less than $M/2$, such misbehavior will have no effect because the final value taken by the requesting agent is the majority value reported the score managers. Two remarks are in order here:

- The score managers are chosen by means of a hash function, that is, randomly. This makes it unlikely for them to collaborate in reporting a common manipulated t_d. If all wrong values can be assumed to be different, malicious score managers could be as numerous as $M - 2$ and there would still be two identical correct values of t_d, which would make malicious behavior ineffectual.
- If the requester can tell which reported reputations are wrong, he can lower the local reputation she awards to the corresponding score managers.

Hence the distributed nature of the algorithm and the redundancy of the computation of the reputation values protect agents against *slandering attacks by the score managers*.

From the discussion above, we see that Protocol 1 is self-enforcing. Also, all agents are guaranteed to get as high a reputation as they fairly deserve, so the protocol is strictly co-utile.

5 Reputation-based Co-utility Enforcement

As discussed in Sect. 2, negative payoffs resulting from helping others (e.g. the bandwidth spent on transferring a file to another peer) may dissuade rational selfish agents from collaborating, thereby preventing a collaborative protocol from being co-utile. Negative payoffs associated to the execution of the s-th protocol step by an agent $\mathcal{P}_i$ can be formally added to the agent's aggregated utility $u_{i,s}$ as a negative cost $o_{i,s}$ weighted by a coefficient α_i^o expressing the importance attached by $\mathcal{P}_i$ to the cost:

$$u'_{i,s} = u_{i,s} - \alpha_i^o \cdot o_{i,s}$$

If, as a result of the weighted negative cost, the aggregated utility $u'_{i,s}$ is negative, the agent will rationally choose not to execute the s-th protocol step.

Reputation can be used as a positive incentive that neutralizes negative payoffs, as long as it is aligned with the nature of the negative payoffs. For example, if costs represent the reluctance of agents to collaborate with each other (because of the fear to interact with strangers or misbehaving agents), reputations can reflect how much a certain agent is trusted based on the outcome of past interactions with her. In this case, positive reputations t_j of other peers $\mathcal{P}_j$ can be used to neutralize the negative costs incurred by a certain agent $\mathcal{P}_i$ when executing a protocol step that would be beneficial for $\mathscr{P}_j$. Thus, $\mathscr{P}_j$'s reputation at the time of running step s, say $t_{j,s}$, is subtracted from the cost, as follows:

$$u'_{i,s} = u_{i,s} - \alpha_i^o \cdot (o_{i,s} - \alpha_i^t \cdot t_{j,s}), \tag{1}$$

where α_i^t is explained next. Putting aside the other atomic utilities, the cost $o_{i,s}$ incurred by agent $\mathscr{P}_i$ can be seen as a threshold that specifies the *minimum reputation* value that $\mathscr{P}_j$ should have in order for $\mathscr{P}_i$ to follow the protocol that will also help $\mathscr{P}_j$. In this context, α_i^t is a coefficient that allows coherently aggregating cost and reputation values (because global reputations are normalized in the range [0..1], whereas the cost can be an unbounded value). In this way, the α_i^o coefficient now weights the difference between the actual cost $o_{i,s}$ and the agent reputation $t_{j,s}$ that compensates the former.

If several different costs (and reputations) can be associated to an action (e.g., in eBay, sellers are rated independently according to delivery speed, shipping costs, communication, etc.), this approach can be generalized in at least two different ways:

- *Aggregation*. For each peer, costs are aggregated into a single cost, and reputations are aggregated into a single reputation, so that the resulting aggregated reputation can compensate the resulting aggregated cost as per Expression (1).
- *Separation*. For each peer, each cost and each reputation are separately managed. In this case, the utility functions for the agents considering to collaborate will incorporate one compensation term for each cost type and corresponding compensating reputation type. That is, instead of subtracting a single term as in Expression (1), one would subtract one term for each cost-reputation pair being considered.

Reputations can also be viewed as atomic utilities that agents wish to increase (because by doing so they also increase the willingness of other agents to collaborate with them). As a result, an agent $\mathscr{P}_i$ may consider as an additional atomic utility her own reputation gain resulting from executing a certain protocol step s:

$$u'_{i,s} = u_{i,s} + \alpha_i^t \cdot (\hat{t}_{i,s+1} - t_{i,s}). \tag{2}$$

In Expression (2), $t_{i,s}$ is the reputation of $\mathscr{P}_i$ before running step s and $\hat{t}_{i,s+1}$ is an estimate of $\mathscr{P}_i$'s reputation $t_{i,s+1}$ after running step s. If agents rate the outcomes of actions in an objective and/or similar way (e.g. spent bandwidth, binary fulfillment of an action, etc.), then the exact value of $t_{i,s+1}$ can be anticipated; otherwise, if reputation values result from subjective opinions (e.g., how pleasant was a trip in a shared car), only the estimate $\hat{t}_{i,s+1}$ can be computed. This shows the importance of agents measuring reputations in a homogeneous way.

With the utility of Expression (2), agents are motivated to increase their reputations indefinitely. However, if reputation values measure the difference between the cost incurred to help other agents and the cost caused to other agents for the help received from them (as in the file sharing scenario), agents are no longer interested in systematically increasing their reputation (because of the incurred costs). They will rather aim at maintaining a reputation value that is barely sufficient to obtain collaboration from other peers (e.g. slightly greater than 0 in the file sharing scenario, which means that the agent has helped others more than what others have helped him). Let such target reputation value be t_{target}. Then the utility expression that represents the interest of the agent in achieving this value (but not more than it) is:

$$u'_{i,s} = u_{i,s} + \alpha_i^t \cdot \min((\hat{t}_{i,s+1} - t_{i,s}), (t_{target} - t_{i,s})). \quad (3)$$

In plain words, Expression (3) reflects that an agent that accumulates more reputation than the target does not obtain any more utility than if she sticks to the target. Hence, if step s of the protocol takes the agent beyond the reputation target, she will not be interested any more in continuing to help others in the next step, because helping is costly and she does not need to increase her reputation further. If the agent subsequently requests and obtains help from other peers, her reputation will decrease; as soon as it falls below the target, the agent will be again interested to help others in order to reach the target once more. Thus, agents with utility given by Expression (3) tend to stick to the target reputation, rather than systematically increasing their reputation. If reputation is the difference between costs borne and costs caused to others, everyone's target reputation will be just slightly above 0, which results in a fair distribution of costs.

The additional atomic utilities considered in the previous expressions can also be combined. For example, an agent may evaluate both the reputation of the peer she is helping (with regard to the cost this help will cause) and also the reputation increase that this action is likely to bring to her own reputation. Depending on the importance attached by each agent to each atomic utility and the values of these utilities/reputations, agents with low reputations may become motivated to collaborate more (in order to increase their reputations), even with agents with not so high reputations. This feature is useful in the file sharing scenario, for example, in which reputations measure the difference between the spent upload and download bandwidths: agents with a low reputation (i.e., those who have downloaded more data than they have uploaded) will be more motivated to fulfill file requests by other peers; otherwise, agents with good reputation are likely to turn down their file requests (because of the low reputation of the requesters).

The reputation calculation protocol enables all agents in the network to faithfully demonstrate their reputation/collaboration record in an anonymous way. Thus, the global reputation t_i achieved by each agent (which can influence the decision of other peers to collaborate) is not a component of the secret type of an agent (which could be manipulated to cheat other peers into changing their strategies), but rather it is public global information that is both accessible and reliably verifiable by the other peers at all times. This fact, in turn, ensures that the reputation atomic utility is a reliable incentive to turn a non-co-utile protocol into a co-utile one.

6 Conclusions

Co-utility happens when selfish agents rationally help each other. However, in many situations, negative payoffs (cost, reluctance, fear, etc.) may render mutual help not rational. In this chapter, we have demonstrated the use of reputation to provide artificial incentives that can compensate negative payoffs, thereby making co-utility still attainable. Specifically, we have described the adaptation and extension the EigenTrust mechanism to obtain a general distributed reputation management protocol that can be applied to a variety of scenarios and reputation needs. The resulting protocol is itself strictly co-utile, robust against a number of attacks and compatible with peer anonymity. In this way, we have shown how we can achieve a virtuous circle by which the reputation mechanism is self-enforcing and, in turn, enables co-utility in protocols that would not be co-utile without reputation, due to negative payoffs. This provides a key tool to design self-enforcing protocols that favor mutual help and boost social welfare even in case the agents' natural interests are inauspicious to this end.

Acknowledgements Funding by the Templeton World Charity Foundation (grant TWCF0095/AB60 "CO-UTILITY") is gratefully acknowledged. Also, partial support to this work has been received from the Government of Catalonia (ICREA Acadèmia Prize to J. Domingo-Ferrer and grant 2014 SGR 537), the Spanish Government (projects TIN2014-57364-C2-1-R "SmartGlacis", TIN2015-70054-REDC and TIN2016-80250-R "Sec-MCloud") and the European Commission (projects H2020-644024 "CLARUS" and H2020-700540 "CANVAS"). The authors are with the UNESCO Chair in Data Privacy, but the views in this work are the authors' own and are not necessarily shared by UNESCO or any of the funding bodies.

References

1. Adler, B., de Alfaro, L.: A content-driven reputation system for the Wikipedia. In: Proceedings of the 16th International Conference on World Wide Web (WWW). pp. 261–270. ACM, New York (2007)
2. BlaBlaCar: http://www.blablacar.com/. Accessed 14 Oct 2015
3. Carverlee, J., Liu, L., Webb, S.: The SocialTrust framework for trusted social information management: architecture and algorithms. Inf. Sci. **180**(1), 95–112 (2010)
4. Domingo-Ferrer, J., Megías, D.: Co-utility for digital content protection and digital forgetting. In: Proceedings of the 15th Annual Mediterranean Ad Hoc Networking Workshop, MedHocNet 2016, pp. 1–7. IEEE, New York (2016)
5. Domingo-Ferrer, J., Farràs, O., Martínez, S., Sánchez, D., Soria-Comas, J.: Self-enforcing protocols via co-utile reputation management. Inf. Sci. **367–368**, 159–175 (2016)
6. Domingo-Ferrer, J., González-Nicolás, Ú.: Rational behavior in peer-to-peer profile obfuscation for anonymous keyword search. Inf. Sci. **185**, 191–204 (2012)
7. Domingo-Ferrer, J., Martínez, S., Sánchez, D., Soria-Comas, J.: Co-utility: self-enforcing protocols for the mutual benefit of participants. Eng. Appl. Artif. Intell. **59**, 148–158 (2017)
8. Domingo-Ferrer, J., Sánchez, D., Soria-Comas, J.: Co-utility: self-enforcing collaborative protocols with mutual help. Prog. Artif. Intell. **5**(2), 105–110 (2016)
9. Domingo-Ferrer, J., Soria-Comas, J., Ciobotaru, O.: Co-utility: self-enforcing protocols without coordination mechanisms. In: Proceedings of the 2015 International Conference on In-

dustrial Engineering and Operations Management, IEOM 2015, pp. 1–17. IEEE, New York (2016)
10. Feldman, M., Padimitriou, C., Chuang, J., Stoica, I.: Free-riding and whitewashing in peer-to-peer systems. IEEE J. Sel. Areas Commun. **24**(5), 1010–1019 (2006)
11. Guha, R., Kumar, R., Raghavan, P., Tomkins, A.: Propagation of trust and distrust. In: Proceedings of the 13th International Conference on the World Wide Web, pp. 403–412. ACM, New York (2004)
12. Hoffman, K., Zage, D., Nita-Rotaru, C.: A survey of attack and defense techniques for reputation systems. ACM Comput. Surv. **42**(1):art 1 (2009)
13. Kamvar, S.D., Schlosser, M.T., Garcia-Molina, H.: The EigenTrust algorithm for reputation management in P2P networks. In: Proceedings of the 12th International Conference on World Wide Web, pp. 640–651. ACM, New York (2003)
14. Kim, Y.A., Phalak, R.: A trust prediction framework in rating-based experience sharing social networks without a Web of Trust. Inf. Sci. **191**, 128–145 (2012)
15. Levine, B.N., Shields, C., Margolin, N.B.: A survey of solutions to the Sybil attack. Tech report, pp. 2006–052. University of Massachusetts Amherst, Amherst, MA (2006)
16. Marti, S., Garcia-Molina, H.: Taxonomy of trust: categorizing P2P reputation systems. Comput. Netw. **50**(4), 472–484 (2006)
17. Pouwelse, J.A., Garbacki, P., Epema, D.H.J., Sips, H.J.: The Bittorrent P2P file-sharing system: measurements and analysis. In: 4th International Workshop on Peer-To-Peer Systems, LNCS 3640, pp. 205–216. Springer, Berlin (2005)
18. Sánchez, D., Martínez, S., Domingo-Ferrer, J.: Co-utile P2P ridesharing via decentralization and reputation management. Transp. Res. Part C **73**, 147–166 (2016)
19. Singh, A., Liu, L.: TrustMe: anonymous management of trust relationships in decentralized P2P systems. In: Proceedings of the 3rd International Conference on Peer-to-Peer Computing, pp. 142–149 (2003)
20. Srivatsa, M., Xiong, L., Liu, L.: TrustGuard: countering vulnerabilities in reputation management for decentralized overlay networks. In: Proceedings of the 14th International Conference on the World Wide Web, pp. 422–431. ACM, New York (2005)
21. Soria-Comas, J., Domingo-Ferrer, J.: Co-utile Collaborative Anonymization of Microdata. In: Torra, V., Narukawa, Y. (eds.) MDAI, pp. 192–206. Springer, Berlin (2015)
22. Turi, A.N., Domingo-Ferrer, J., Sánchez, D., Osmani D.: Co-utility: conciliating individual freedom on common good in the crowd based business model. In: Proceedings of the 12th International Conference on e-Business Engineering, ICBE 2015, pp. 62–67. IEEE, New York (2015)
23. Turi, A.N., Domingo-Ferrer, J., Sánchez, D., Osmani, D.: A co-utility approach to the mesh economy: the crowd-based business model. Review of Managerial Science. (2016, in press)
24. Turi, A. N., Domingo-Ferrer, J., Sánchez, D.: Filtering P2P loans on co-utile reputation. In: Proceedings of the 13th International Conference on Applied Computing-AC 2016, pp. 139–146 (2016)
25. Walsh, K., Sirer, E.G.: Experience with and object reputation system for peer-to-peer filesharing. In: Symposium on Networked System Design and Implementation-NSDI'06 (2006)
26. Zhou, R., Hwang, K.: PowerTrust: a robust and scalable reputation system for trusted peer-to-peer computing. IEEE Trans. Parallel Distrib. Syst. **18**(4), 460–473 (2007)
27. Zhou, R., Hwang, K., Cai, M.: GossipTrust for fast reputation aggregation in peer-to-peer networks. IEEE Trans. Knowl. Data Eng. **20**(9), 1282–1295 (2008)

On the Different Forms of Individual and Group Strategic Behavior, and Their Impact on Efficiency

Salvador Barberà, Dolors Berga and Bernardo Moreno

Abstract We survey a number of results regarding incentives and efficiency that have been recently added to the social choice literature, and we establish parallels and differences with the concepts and results obtained so far by co-utility theory. Our main purpose is to facilitate the convergence between researchers who end up dealing with similar issues in response to rather different motivations and backgrounds.

1 Introduction

This paper is an attempt to approximate the work of social choice theorists to that of computer scientists, prompted by the emergence of the attractive concept of co-utility (see [10, 11]).

Although surprising some decades ago, the connection between the economists' approach to mechanism design and the concerns of computer scientists is now a well-established fact, even if there are still communication barriers based on mutual ignorance and differences in language and motivation. But the appearance of new fields like that of computational mechanism design (see, for example, [17]) witnesses the tendency to converge between disciplines that have bumped into similar problems from rather different starting points. We, the authors of the present essay, are economists working in the area of mechanism design, and mostly within the subfield of social choice. And, again, there is a growing literature on computational social

S. Barberà (✉)
MOVE, Universitat Autònoma de Barcelona, and Barcelona GSE, Edifici B, 08193 Bellaterra, Catalonia, Spain
e-mail: salvador.barbera@uab.cat

D. Berga
Departament d'Economia, Universitat de Girona, C/Universitat de Girona, 10, 17003 Girona, Catalonia, Spain
e-mail: dolors.berga@udg.edu

B. Moreno
Facultad de Ciencias Económicas y Empresariales, Departamento de Teoría e Historia Económica, Universidad de Málaga, Campus de El Ejido, 29071 Málaga, Spain
e-mail: bernardo@uma.es

J. Domingo-Ferrer and D. Sánchez (eds.), *Co-utility*, Studies in Systems, Decision and Control 110, DOI 10.1007/978-3-319-60234-9_3

choice (see [9] for an introduction) that enriches the interaction between the same two scientific communities. This literature has its own concerns and meeting points, but we are not aware of any previous effort to put together the advances regarding co-utility with results in our area that ring very similar bells. Hence, our purpose is to describe a line of work in social choice theory whose motivation and conclusions we can clearly describe from the economists' point of view, and to comment, much more tentatively, on the analogies and the still existing differences between the work that we survey and the research on co-utility. Although very fragmentary, we hope that this essay will be informative for computer scientists and also useful to economists, as a reminder of how worthy it is to connect our partial knowledge with that of other fields.

Let us spell out some of the limits of this essay. We shall present results whose main interpretation refers to voting systems, to be used for the collective choice of candidates, or the determination of levels of provision of public goods. Most of the issues we raise and of the concepts we introduce can be extended to other types of decisions involving private goods, but for the sake of simplicity we stick to the traditional social choice framework. This makes the comparison with co-utility a bit harder, since in principle that theory does not distinguish between these two subcases, and we explicitly exclude in the present work references to auctions, or to other very central concerns involving private goods like allocation of indivisible goods or matching. But we hope it provides a good enough starting point, and that it makes our presentation specific.

We also acknowledge that we shall not go as far as to provide a complete reconciliation between the results we present and the concepts, techniques and results of co-utility theory. Rather, we will try to point out what we believe to be common concerns, and what we still feel are possible divergences.

The possibility of strategic voting has been discussed since ancient times (see the references to Pliny the Younger, Ramon Llull and later authors in the wonderful book by [14]; a brief summary of that history is provided in [3]). At the early ages of modern social choice theory, it was mentioned but not analyzed by [2]. Early contributors to its study were [12, 18, 19, 23], but the real start of a thorough analysis of manipulation and strategy-proofness in voting came with the works of [13, 22]. Their main result proved that any non-trivial social choice function (or voting rule, depending on the interpretation) is manipulable, as soon as we must choose from more than two alternatives and the preferences of agents are unrestricted. This result generated two large streams of literature. One concentrated on the idea that manipulation per se is nothing but the play of a game, and studied the consequences of having agents behave strategically. Implementation theory emerged from that, in all the richness that derives from considering a variety of possible games and applying to them different solution concepts. The other stream, where we place the results that follow, interpreted the negative result of Gibbard and Satterthwaite as an invitation to concentrate on the specifics of each relevant case, and to characterize those situations where one could expect restrictions on agent's preferences to hold and to allow for the existence of attractive strategy-proof rules. Two such examples are provided by environments where agents' preferences can be assumed to be single-peaked, or others where alternatives consist of collections of objects and preferences satisfy separability conditions. In these environments, and under the preference restrictions that they suggest, one can identify many strategy-proof social choice functions. And this rings a bell regarding co-utility's requirement that individual agents should follow a prescribed protocol. However, one can ask for more, and not all environments admitting strategy-proof rules allow for the additional requirement that groups of agents should also find it beneficial to abide to the rules. Group strategy-proofness, or some intermediate concept qualifying the joint behavior of voters, becomes an additional objective for the social choice theorist, as well as an analysis of the relative efficiency of those rules that satisfy such conditions. Until now co-utility has only demanded individual incentives to be appropriate, and not insisted directly on the strategic behavior of groups. Its concern with efficiency, however, relates to the possibility of avoiding joint manipulations by the set of all players, and is therefore quite connected with the type of group behavior that we shall study here. Hence our interest in exposing what we know as social choice theorists, to suggest how far the analogies with co-utility go, and to express the need we still feel for further approaches.

2 The General Framework

Let A be the set of *alternatives* and $N = \{1, 2, \ldots, n\}$ be the set of *agents* (with $n \geq 2$). Let capital letters $S, T \subset N$ denote subsets of agents while lower case letters s, t denote their cardinality.

Let $\mathscr{R}$ be the set of complete, reflexive, and transitive orderings on A and $\mathscr{R}_i \subseteq \mathscr{R}$ be *the set of admissible preferences for agent* $i \in N$. A *preference profile,* denoted by $R_N = (R_1, \ldots, R_n)$, is an element of $\times_{i \in N} \mathscr{R}_i$. As usual, we denote by P_i and I_i the strict and the indifference part of R_i, respectively. Let $C, S \subset N$ be two coalitions such that $C \subset S$. We will write the subprofile $R_S = (R_C, R_{S \backslash C}) \in \times_{i \in S} \mathscr{R}_i$ when we want to stress the role of coalition C in S. Then the subprofiles $R_C \in \times_{i \in C} \mathscr{R}_i$ and $R_{S \backslash C} \in \times_{i \in S \backslash C} \mathscr{R}_i$ denote the preferences of agents in C and in $S \backslash C$, respectively.

A *social choice function* (or a *rule*) is a function $f : \times_{i \in N} \mathscr{R}_i \to A$.

All along the paper we will focus on rules that are non-manipulable, either by a single agent or by a coalition of agents. We first define what we mean by a manipulation and then we introduce the well known concepts of *strategy-proofness* and *group strategy-proofness*.

Definition 1 A social choice function f is (strongly) group manipulable on $\times_{i \in N} \mathscr{R}_i$ at $R_N \in \times_{i \in N} \mathscr{R}_i$ if there exist a coalition $C \subset N$ and $R'_C \in \times_{i \in C} \mathscr{R}_i$ ($R'_i \neq R_i$ for any $i \in C$) such that $f(R'_C, R_{-C}) P_i f(R_N)$ for all $i \in C$. We say that f is individually manipulable if there exists a possible manipulation where coalition C is a singleton.

Definition 2 A social choice function f is (weakly) group strategy-proof on $\times_{i \in N} \mathscr{R}_i$ if f is not (strongly) group manipulable for any $R_N \in \times_{i \in N} \mathscr{R}_i$. Similarly, f is strategy-proof if it is not individually manipulable.

Definition 3 A social choice function f is (weakly) efficient if for any $R_N \in \times_{i \in N} \mathscr{R}_i$, there is no alternative a such that $a P_i f(R_N)$ for all $i \in N$.

Definition 4 A social choice function f is strongly efficient if for any $R_N \in \times_{i \in N} \mathscr{R}_i$, there is no alternative a such that $a R_i f(R_N)$ for all $i \in N$ and $a P_j f(R_N)$ for some $j \in N$.

Note that (weak) group strategy-proofness implies (weak) efficiency but the converse does not hold. For example, in the first application defined in Sect. 3, the rule choosing the mean of the peaks is an efficient rule, but it is manipulable.

In our definitions above, and at different points in the text, we qualify different notions as being weak or strong depending on whether strict domination occurs for all agents involved, or only for some. Since most of the work we are going to discuss assumes for simplicity that all agents have strict preferences, the qualification will not in general be needed here, although it is important in other contexts, as those where private goods are present, where it becomes quite essential. Another qualification we want to make regards the definition of efficient rules. Typically we consider social choice functions whose range covers the whole set of alternatives. When applied to rules that are not onto, the efficiency notion that we apply restricts attention to those alternatives that are in the range.

3 Possibility and Impossibility Results in Designing Strategy-Proof Social Choice Rules

We first present Gibbard and Satterthwaite's on the impossibility of designing non-trivial strategy-proof rules when individual preferences are non-restricted. After that, we present three results characterizing classes of strategy-proof rules that may be defined in different settings where agents' preferences are appropriately restricted given the type of problem under consideration. These results will provide reference points to discuss the connections between incentives and efficiency in the following section.

Definition 5 A social choice function is dictatorial if there exists an agent $d \in N$ such that for any preference profile $R_N \in \mathscr{R}^n$, $f(R)$ is such that $f(R_N) R_d x, \forall x \in A$.

Theorem 1 [13, 22] *Any social choice function f on $\mathscr{R}^n$ with at least three alternatives in the range is either manipulable or dictatorial.*

If we admit that dictatorial rules are trivial, and we also consider as such those that only select among two alternatives, the result expresses the impossibility of designing non-trivial rules operating on the universal domain of preferences.

But then, starting from this impossibility, one may ask whether there exist environments where it is natural to expect agents' preferences to be restricted, and where, given these restrictions, strategy-proof rules may be defined within these natural settings. Much work has been done in characterizing restricted domains of preferences where attractive and strategy-proof rules can be found. We present three cases where such environments have been identified and strategy-proof rules characterized: (1) the choice of candidates or levels of provision of pure public goods when agents evaluate alternatives in one dimension and exhibit single-peaked preferences, (2) the choice of multiple candidates to join a club, when voters' preferences are separable, and (3) the choice of levels for different characteristics of candidates, where these can be seen as points in a grid, and the preferences of voters satisfy a generalized version of single-peakedness and separability. The latter case subsumes the preceding two, but presenting each case in turn will facilitate our exposition.

3.1 One Dimensional Decisions Under Single-Peaked Preferences

Let A be an interval in $\mathbb{R}$, and consider a linear order of A, say $>$. We now define the concept of a single-peaked preference on A (given $>$).

Definition 6 A preference $R_i \in \mathscr{R}_i$ is **single-peaked** on A if (1) there exists $\tau(R_i) \in A$ such that $\tau(R_i) P_i y$ for any $y \in A$, and (2) for any $y, z \in A$, $[z < y < \tau(R_i)$ or $z > y > \tau(R_i)]$ then $y P_i z$.

Let $\mathscr{SP}$ be the set of individual single-peaked preferences on A

Definition 7 A social choice function $f : \mathscr{SP}^n \to A$ is a generalized Condorcet-winner rule if, for any profile $R_N \in \mathscr{SP}^n$, $f(R_N) = median\{\tau(R_1), \ldots, \tau(R_n), p^1, \ldots, p^{n-1}\}$, where $P(f) = \{p^1, \ldots, p^{n-1}\}$ is a list of parameters in A.

We will consider rules satisfying the additional property of anonymity. Formally,

Definition 8 A social choice function $f : \mathscr{SP}^n \to A$ is anonymous if for any $R_N \in \mathscr{SP}^n$ and for any permutation of agents $\sigma : N \to N$, $f(R_N) = f(R_{\sigma(N)})$.

That is, anonymity says that the name of the agents involved in the social decision does not matter; only the preference profile matters, not the particular opinion of an individual.

Theorem 2 [16] *A social choice function f on $\mathscr{SP}^n$ is anonymous, onto, and strategy-proof if and only if it is a generalized Condorcet-winner rule.*

This is not the most general characterization result in the single-peaked context, but we add the condition of anonymity for simplicity and because anonymous rules are of special interest. The reader may appreciate that choosing the median voter's best alternative is a special case of those rules that satisfy strategy-proofness in the single-peaked domain.

3.2 Choosing Multiple Candidates for a Club

Let $A = 2^{\mathcal{O}}$, where $\mathcal{O} = \{o_1, \ldots, o_K\}$ is a finite set of objects. Let individual preferences be linear orders on $2^{\mathcal{O}}$ (including the empty set). Define the set of "good" objects as $G(K, R) = \{o_k \in \mathcal{O} : \{o_k\} P \varnothing\}$, and the set of "bad" objects as $K \backslash G(K, R) = \{o_k \in \mathcal{O} : \varnothing P \{o_k\}\}$.

Definition 9 R is an individual separable preference on $2^{\mathcal{O}}$ if and only if for any set T and any object $o_l \notin T$, $T \cup \{o_l\} P T$ if $o_l \in G(K, R)$.

Let $\mathcal{S}$ denote the set of all separable preferences.

Definition 10 Let $q \in \{1, \ldots, n\}$. The social choice function f on $\mathcal{S}^n$ defined so that for any $R_N \in \mathcal{S}^n$,

$$f(R_N) = \{o_k \in \mathcal{O} : |\{i : o_k \in G(K, R_i)\}| \geq q\}$$

is called voting by quota q.

These rules turn to be important because they have been obtained as the unique class satisfying desirable properties in some problems. One of these properties is neutrality.

Definition 11 A social choice function $f : \mathcal{S}^n \to 2^{\mathcal{O}}$ is neutral if, for any $R_N \in \mathcal{S}^n$ and for any permutation of objects $\mu : \mathcal{O} \to \mathcal{O}$, $f(\mu(R_1), \mu(R_2), \ldots, \mu(R_n)) = \mu(f(R_N))$, where for each $j \in N$, $\mu(R_j)$ is the preference obtained from R_j by permuting the objects according to μ.

Neutrality is a general property that social choice functions may or may not satisfy. In general terms, it requires that all alternatives should be treated equally by the rule, and that their labelling is irrelevant. We use it here for simplicity, although more general results are also known. Notice that, in the case we consider here, alternatives are sets of objects, hence our notation.

Theorem 3 [8] *A social choice function f on $\mathcal{S}^n$ is anonymous, neutral, onto, and strategy-proof if and only if it is a voting by quota rule.*

Notice again that this is not the most general characterization theorem for strategy-proof rules in that environment, but that restricting attention to neutral and anonymous rules allows us to point at particularly simple ones: those where each agent votes for those candidates that she deems good, and all candidates who get at least q votes are elected.

3.3 Choosing from a Grid

Let $\mathscr{K} = \{1, \ldots, K\}$ be a finite set of $K \geq 2$ coordinates, and $B_k = [a_k, b_k], a_k < b_k$ be an integer interval of possible values for each of the coordinates. Then, $A = \prod_{k=1}^{K} B_k$ is the set of alternatives which are K-dimensional vectors. We endow A with the L_1-norm, that is, for any $x \in B$, $\|x\| = \sum_{k=1}^{K} | x_k |$. Given $x, y \in B$, the *minimal box containing* x *and* y is defined by

$$MB(x, y) = \{z \in B : \|x - y\| = \|x - z\| + \|z - y\|\}.$$

Definition 12 A preference $R_i \in \mathscr{R}_i$ is *(multidimensional)* **single-peaked** if (1) there exists $\tau(R_i) \in A$ such that $\tau(R_i) P_i y$ for any $y \in A$, and (2) for any $z, y \in B$, if $y \in MB(z, \tau(R_i))$ then $y R_i z$

Let $\mathscr{S} \subseteq \mathscr{R}_i$ be the set of (multidimensional) single-peaked preferences on A. Define $\tau(R_i) = (\tau_1(R_i), \ldots, \tau_K(R_i)) \in A$, where $\tau_k(R_i)$ is the best (or top) alternative of R_i in dimension k.

As announced, notice that this environment includes the two previous ones as special cases, where only one dimension is allowed, or else only two possible values are possible for each dimension. In fact, the following definition encompasses the characteristics of the two types of mechanisms that we have characterized in the previous results.

Definition 13 A social choice function $f : \mathscr{S}^n \to A$, $f = (f_1, \ldots, f_K)$ is a (multidimensional) generalized Condorcet-winner rule if for any $R_N \in \mathscr{S}^n$, $k \in \mathscr{K}$, $f_k(R_N) = median\{\tau_k(R_1), \ldots, \tau_k(R_n), p_k^1, \ldots, p_k^{n-1}\}$, where $P_k(f) = \{p_k^1, \ldots, p_k^{n-1}\}$ is a list of parameters in B_k.

Note that in the one-dimensional case, that is for $K = 1$, these rules precipitate to the rules in Definition 7.

Theorem 4 [7] *A social choice function f on $\mathscr{S}^n$ is anonymous, onto, and strategy-proof if and only if it is a generalized Condorcet-winner rule.*

We close this Section by emphasizing that, while the rules we have presented are all strategy-proof, they differ very much regarding their efficiency, for the case in which agents vote straightforwardly. The generalized Condorcet-winner rules are efficient in the one-dimensional case. What is more: not only they are immune to profitable deviations by the set of all agents, but also to profitable deviations by groups of agents: In the language we have already presented in Sect. 2, they are group strategy-proof, which implies they are efficient.

The following example shows that voting by quota q is an inefficient rule in the separable domain, and a fortiori, that multidimensional generalized Condorcet-winner also fail to satisfy efficiency.

Example 1 Let us use quota 2 for electing subsets of a three candidate election with three voters, in a separable environment. Let $N = \{1, 2, 3\}$, $K = 3$ and consider voting by quota 2. Let R_N be as follows: the preferences of agent 1 are such that $\tau(R_1) = \{o_1, o_2\}$ and $\varnothing P_1 \{o_1, o_2, o_3\}$, the preferences of agent 2 are such that $\tau(R_2) = \{o_2, o_3\}$ and $\varnothing P_2 \{o_1, o_2, o_3\}$, and the preferences of agent 3 are such that $\tau(R_3) = \{o_1, o_3\}$ and $\varnothing P_3 \{o_1, o_2, o_3\}$. Observe that $f(R_N) = \{o_1, o_2, o_3\}$. This outcome is clearly inefficient.

The reader will notice that the example can be generalized for any possible quota q, any $K \geq 2$ and any number of agents.

4 Equivalence Between Individual and Group Strategy-Proofness

It is clear from their definitions that group implies individual strategy-proofness, but not the other way around. Yet in many interesting applications, once we have a function that satisfies the weak requirement, it also meets the stronger one. For other equally interesting applications, the equivalence does not hold. Let us provide some examples. In public goods settings, the ones we shall deal with in this essay, generalized Condorcet-winner rules are both individual and group strategy-proof under single-peaked preferences and one dimensional spaces. Both properties also hold for properly chosen social choice functions in one dimensional cases when preferences are single-dipped. And although we shall not present these results here, the equivalence between individual and group strategy-proofness also holds for interesting classes of social choice functions defined in contexts with private goods. This is the case, for example, for settings involving matching, rationing, house allocation cost sharing or simple auctions.

In contrast, other interesting environments admitting strategy-proof mechanisms allow for these mechanisms to be manipulable by groups. These include two of the examples we present here for the case of public goods, involving multi-dimensional choices, and, in the case of environments with private goods and quasi-linear preferences, the well-known family of Vickrey-Clarke-Groves mechanisms. In fact, in our examples group strategy-proofness is violated in a strong manner, and the only rules that are strategy-proof do not even satisfy efficiency.

In view of this dichotomy, we have investigated whether the equivalence, when it holds, needs a case-by-case explanation, or is the result of some common features underlying the different models. In what follows, we present a result that identifies such common features for the case of public goods, and proves that they precipitate the equivalence. A similar result also holds for many apparently diverse environments in private goods settings [5].

We end this section by stating the unifying result in our simple public goods setting. In the following section we shall turn attention to the properties of those rules that are strategy-proof but hold that property in environments where group manipulation remains a possibility.

Let $R_N \in \times_{i \in N} \mathscr{R}_i$, and y, z be a pair of alternatives. Denote by $S(R_N; y, z) \equiv \{i \in N : yP_i z\}$, that is, the set of agents who strictly prefer y to z according to their individual preferences in R_N. Moreover, define the concept of lower contour set. For any $x \in A$ and $R_i \in \mathscr{R}_i$, define *the lower contour set of* R_i *at* x as $L(R_i, x) = \{y \in A : xR_i y\}$. Similarly, the strict lower contour set at x is $\overline{L}(R_i, x) = \{y \in A : xP_i y\}$.

Definition 14 Given a preference profile $R_N \in \times_{i \in N} \mathscr{R}_i$ and a pair of alternatives $y, z \in A$, we define a binary relation $\succsim (R_N; y, z)$ on $S(R_N; y, z)$ as follows:

$$i \succsim (R_N; y, z) j \text{ if } L(R_i, z) \subseteq \overline{L}(R_j, y)$$

Definition 15 A preference profile $R_N \in \times_{i \in N} \mathscr{R}_i$ satisfies sequential inclusion for $y, z \in A$ if the binary relation $\succsim (R_N; y, z)$ on $S(R_N; y, z)$ is complete and acyclic.

In [4] we prove the following equivalence result.

Theorem 5 *Let* $\times_{i \in N} \mathscr{R}_i$ *be a domain such that any preference profile satisfies sequential inclusion. Then, any strategy-proof rule* f *on that domain is also group strategy-proof.*

Admittedly, sequential inclusion is a strong condition. Our emphasis here is on the fact that, however restrictive, it identifies environments and social choice rules for which the objectives of co-utility are met in a very strong sense. Incentives for individuals and for groups are perfectly aligned, and efficiency is attained.

While Theorem 5 qualifies the quite generalized belief that incentive compatibility and efficiency are always incompatible, the tension between these two objectives remains in other contexts. In the next section we study some of the consequences that derive from operating under strategy-proof mechanisms that are not immune to group manipulations.

5 A Weaker Notion to Control Manipulation by Groups

The previous characterization results have proven that there is a large gap between those environments in which individual strategy-proofness implies group strategy-proofness and those where this equivalence does not hold. In the first case, a reconciliation between the two main objectives of co-utility is automatic, while in the second case it is problematic, to say the least.

We shall follow here [6] and study conditions that are intermediate between individual and group strategy-proofness. That will allow us to classify the strategy-proof rules in the multidimensional case into two different classes, according to the credibility that one can attach to the manipulative actions of groups.

Definition 16 Let f be a social choice function on $\times_{i\in N}\mathscr{R}_i$. Let $R_N \in \times_{i\in N}\mathscr{R}_i$ and $C \subseteq N$. A subprofile $R'_C \in \times_{i\in C}\mathscr{R}_i$ such that $R'_i \neq R_i$ $\forall i \in C$ is a **profitable deviation** of coalition C against profile R_N if $f(R'_C, R_{N\setminus C})P_i f(R_N)$, for any agent $i \in C$.

Definition 17 Let f be a social choice function on $\times_{i\in N}\mathscr{R}_i$. Let $R_N \in \times_{i\in N}\mathscr{R}_i$ and $C \subseteq N$. We say that $R'_C \in \times_{i\in C}\mathscr{R}_i$ a profitable deviation of C against R_N is **credible** if for all $i \in C$ and all $\overline{R}_i \in \mathscr{R}_i$, then $f(R'_C, R_{N|C})R_i f(\overline{R}_i, R'_{C\setminus\{i\}}, R_{N\setminus C})$.

On other terms, a profitable deviation by C from $R_N = (R_C, R_{N\setminus C})$ is credible if R'_C is a Nash equilibrium of the game among agents in C, when these agents strategies are their admissible preferences and the outcome function is $f(\cdot, R_{N\setminus C})$.

Definition 18 A social choice function f on $\times_{i\in N}\mathscr{R}_i$ is **immune to credible deviations** if for any $R_N \in \times_{i\in N}\mathscr{R}_i$ and $C \subseteq N$, there is no credible profitable deviation of C against R_N (that is, for any profitable deviation $R'_C \in \times_{i\in C}\mathscr{R}_i$ of C against R_N, there exists an agent $i \in C$ such that $f(\overline{R}_i, R'_{C\setminus\{i\}}, R_{N\setminus C})P_i f(R'_C, R_{N|C})$ for some $\overline{R}_i \in \mathscr{R}_i$).

Although formulated in a different manner, this concept of immunity turns out to be equivalent, in the contexts that we describe, to the notion of strong coalition-proofness proposed by Peleg and Sudhölter [21]. In their case, they described the concept of strong coalition proof equilibrium and then proposed that a social choice function be required to have truthful revelation as an equilibrium of their kind (see also Peleg [20]).

Note that immunity implies strategy-proofness (assuming the existence of the best profitable deviations for single agents). However the converse does not hold in general.

Lemma 1 *Any social choice function f on $\times_{i\in N}\mathscr{R}_i$ that is immune to credible deviations is strategy-proof.*

By definition we can observe that any generalized Condorcet-winner rule on the single-peaked domain is both (group) strategy-proof and immune to credible deviations. Moreover, any voting by quota q is strategy-proof on the separable domain but only voting by quota 1 and n are immune to credible deviations, as we proved in [6]. On the (multidimensional) single-peaked domain, any multidimensional generalized Condorcet-winner is strategy-proof but only those whose list of parameters are degenerate in at least $K - 1$ dimensions are immune to credible deviations, as we also proved in the mentioned paper.[1]

In the case of choosing candidates we have the following results.

Proposition 1 *Let $n > 3$. Then, voting by quota 1 and n are the only voting by quota rules satisfying immunity to credible deviations.*[2]

[1] A list of parameters is degenerate if they are all equal.

[2] The case $n = 2$ gives the same results as $n > 3$. The case $n = 3$ requires special treatment. We obtain that when $n = 3$ and $K = 2$, any voting by quota rule is immune to credible deviations. When $n = 3$ and $K \geq 3$, voting by quota 1 and 3 are the only voting by quota rules satisfying immunity to credible deviations.

In the case of choosing from a grid, here are the results.

Proposition 2 *Let $n > 3$. Let f be a generalized Condorcet-winner rule. If f is defined by lists of parameters that are non-degenerate in at least two dimensions, then f is not immune to credible deviations.*

Proposition 3 *Let $n \geq 2$. Let f be a generalized Condorcet-winner rule. If f is defined by lists of parameters that are degenerate in at least $K - 1$ dimensions, then f is immune to credible deviations.*[3]

Since the notion of credibility may have different expressions, we shall offer now some alternatives to our main definition, in addition to the already mentioned possibility of reformulating it in terms of strong coalition-proof equilibria.

The first variant will be one where, instead of letting agents in C to have any choice of preferences as a strategy, we restrict them to either use strategy R'_i or to revert to strategy R_i. The resulting notion of a credible deviation will be stronger than ours. However, as shown below the set of rules that are immune to credible deviations will be the same (after a minimal qualification) under either definition.

Definition 19 Let f be a social choice function on $\times_{i \in N} \mathscr{R}_i$. Let $R_N \in \times_{i \in N} \mathscr{R}_i$ and $C \subseteq N$. We say that $R'_C \in \times_{i \in C} \mathscr{R}_i$ a profitable deviation of C against R_N is (type 1) **credible** if $f(R'_C, R_{N|C}) R_i f(R_i, R'_{C \setminus \{i\}}, R_{N \setminus C})$ for all $i \in C$. A social choice function f on $\times_{i \in N} \mathscr{R}_i$ is **immune to** (type 1) **credible deviations** if for any $R_N \in \times_{i \in N} \mathscr{R}_i$, any $C \subseteq N$, there is no (type 1) credible profitable deviation of C against R_N.

Proposition 4 *A social choice function f on $\times_{i \in N} \mathscr{R}_i$ is immune to credible deviations if and only if it is immune to (type 1) credible deviations.*

A second variant will require that in order to be (extensively) credible, deviation R'_C should be a Nash equilibrium for the game where all agents can play any preference, f being the outcome function. If f is strategy-proof, then again the set of immune rules will still be the same under either definition. Otherwise, the equivalence is not true.

Definition 20 Let f be a social choice function on $\times_{i \in N} \mathscr{R}_i$. Let $R_N \in \times_{i \in N} \mathscr{R}_i$ and $C \subseteq N$. We say that $R'_C \in \times_{i \in C} \mathscr{R}_i$ a profitable deviation of C against R_N is (type 2) **credible** if $f(R'_C, R_{N|C}) R_i f(\overline{R}_i, R'_{C \setminus \{i\}}, R_{N \setminus (C \cup \{i\})})$ for all $i \in N$ and all $\overline{R}_i \in \mathscr{R}_i$. A social choice function f on $\times_{i \in N} \mathscr{R}_i$ is **immune to** (type 2) **credible deviations** if for any $R_N \in \times_{i \in N} \mathscr{R}_i$, any $C \subseteq N$, there is no (type 2) credible profitable deviation of C against R_N.

Proposition 5 *Any strategy-proof social choice function f on $\times_{i \in N} \mathscr{R}_i$ is immune to credible deviations if and only if it is immune to (type 2) credible deviations.*

---

[3]The case $n = 3$ requires special treatment and the result obtained is the following: Any generalized Condorcet winner rule defined by lists of parameters such that are non-degenerate in two dimensions is immune to credible deviations. Any generalized Condorcet winner rule defined by non-degenerate lists of parameters in at least three dimensions is not immune to credible deviations.

A third variant of our definition of credibility would result from simply changing our original one, but ask the deviation to be a strong Nash, rather than a Nash equilibrium. The rationale for such proposal would be to allow for several agents to coordinate when defecting from the agreed upon joint manipulation.

Definition 21 Let f be a social choice function on $\times_{i \in N} \mathscr{R}_i$. Let $R_N \in \times_{i \in N} \mathscr{R}_i$ and $C \subseteq N$. We say that $R'_C \in \mathscr{U}^c$ a profitable deviation of C against R_N is **strongly credible** if $f(R'_C, R_{N|C}) R_i f(\overline{R}_S, R'_{C \setminus S}, R_{N \setminus C})$ for all $S \subseteq C$, for all $\overline{R}_S \in \times_{i \in S} \mathscr{R}_i$ and for some $i \in S$. A social choice function f on $\times_{i \in C} \mathscr{R}_i$ is **immune to strongly credible deviations** if for any $R_N \in \times_{i \in C} \mathscr{R}_i$, any $C \subseteq N$, there is no strongly credible profitable deviation of C against R_N.

Proposition 6 *All generalized Condorcet-winner rules are immune to strongly credible deviations.*

After this digression showing the connections between different possible definitions of credibility and their consequences, the following comments refer to our main definition of immunity, as presented in Definition 17.

The distinction between those social choice functions which are immune to credible deviations and those that are not is very relevant for our purpose of connecting our results with those in co-utility theory. In some sense, one may think that the most attractive rules are those that are not only strategy-proof but also immune to credible deviations. If we take as an example the choice of candidates to a club under separable preferences, and we push this point of view, then quota one or quota n would stand out as being especially attractive. However, let us turn the argument around and ask what would be the performance of the rules based on voting by quota q, for values of q higher than one and lower than n.

In that case, we know that there will be some cases where groups of agents can credibly coalesce for their joint profit.

Let us see what may happen in a simple case, which can be easily extended to other cases.

Example 2 (Example 1, continued) Let us use quota 2 for electing subsets of a three candidate election with three voters, in a separable environment. When the voter's preferences are R_N as in Example 1: that is, agent 1 likes o_1 and o_2 but hates o_3, 2 likes o_2 and o_3 but hates o_1, 3 likes o_1 and o_3 but hates o_2, and all of them prefer not having any candidate to have all of them, the outcome from sincere voting would be all candidates, while in fact all agents would prefer no candidate being elected. Hence, the rule clearly yields inefficient outcomes. One way in which the three voters could depart beneficially from sincere voting would be by none of them giving support to any candidate. Their unanimous vote for the empty set would not only give them a result that Pareto improves upon the outcome resulting from sincere voting, but is credible, because no agent could deviate from the agreed upon joint voting strategy and gain from it.

As we already observed, the outcome of voting by quota 2 under the example's preferences is inefficient, because all agents prefer the empty set to the set with all three alternatives which would result from truthful preference revelation. We now remark that their unanimous vote for the empty set would not only give them a Pareto superior result, but also constitute a credible deviation from the truth, since no agent alone could profitably deviate from that joint strategy profile, which is a Nash equilibrium.

The example suggests that, if we select rules that are credibly manipulable, and allow groups of agents that have credible joint strategies to use them, the induced connection between individual preferences and final outcomes, after manipulation, define an efficient social choice function.

Moreover, notice that none of the agents can be distinguished from the others by observing the votes, and that none of the differential features of the voters are revealed.

Hence, we can generate a result that is very close to the desiderata of co-utility, as we shall comment more extensively in the conclusions below.

Here is a more general argument and construct in the same spirit. Take a rule f that is inefficient and admits credible profitable deviations by the grand coalition. Look at the rule f' defined as follows. For those profiles P where no group has a credible profitable deviation, let $f(P) = f'(P)$. Now, for those profiles where there are one or more groups that could credibly profitably deviate, choose a specific one among those that involve the largest number of deviants, and if necessary pick it in such a way that it gives maximal gains to the agent with the smallest indicator, in lexicographic order. Then, let the value of f' be the resulting one from applying the function f to the profile where the agents in the coalition have profitably deviated, while the rest of agents still vote truthfully. Notice that the resulting function is efficient.

6 Conclusions

In this section we comment on the analogies and the differences that we find between our results on the manipulability of social choice functions and the purposes and findings in the of co-utility context.

First of all, there is the obvious analogy that in both cases we are trying to devise a mechanism, represented as a game form that gives rise to a game for each specification of the agents' objectives.

We concentrate on the case of public decisions taken without the possibility of money compensation. Although this decision excludes important parts of the literature on mechanism design, like auction theory, it has the advantage of showing that the issues we are dealing with arise in even the most basic settings.

The two basic notions underlying the design of co-utile protocols can be qualified in several ways. The idea of self-enforcement will depend on the type of game we assume agents will play, and on the associated notion of equilibrium. Our results are expressed for simultaneous games, where protocols are very simple: all agents are required to submit their information at the same time in one shot. While this formulation may be less realistic than formulating an extensive form game involving sequential protocols, it also has its advantages. The subgame perfect equilibria of the latter are very sensitive to the order of play, while in our case we can identify environments where all agents can be endowed with dominant strategies, which in our view is a very satisfactory expression of self enforcement.

The use of strategy-proof mechanisms in extensive form is under study by an emerging literature on the concept of obvious strategy-proofness, defined by [15]. The paper [1] on that subject is particularly close to our present concerns, since it applies to the type of environments we have restricted attention to. We have not been able yet to establish the full connections between that work and co-utility, but we want to call attention toward these very recent papers.

Now, regarding efficiency, there are also possible qualifications, but let us say that, in those cases where we can prove that strategy-proofness implies group strategy-proofness, we can nicely tie the two essential requirements of co-utility into one. This is because group strategy-proofness implies efficiency, when applied to the set of all agents. This is what happens when preferences are restricted to be unidimensional single-peaked.

The more interesting connections arise when we analyze those contexts where strategy-proof rules are not efficient, hence not group strategy-proof, as it happens in the case of multi-dimensional single-peaked environments, and in the more specific case of separable preferences. In that case, we have provided a characterization of those anonymous strategy-proof rules that are immune to profitable deviations, and those that, although strategy-proof, are not immune.

Now, from the point of view of mechanism design, one could claim that the interesting case is the one where agents have no incentive to deviate from the truth, and cannot from profitable coalitions either, since these rules would be strongly self enforceable: no gains could be obtained by not abiding to the rules. But the interesting twist, and the one that approaches our results to co-utility, comes from the opposite point of view. Consider a rule that is not immune to credible deviations among those that we have characterized: for instance take Example 1, the quota two rule for electing subsets of a three candidate election with three voters, in a separable environment. We know that the whole set of agents will be able to credibly manipulate in some cases: Their joint manipulation will take the form of a transaction, where every voter stops supporting some desirable candidates if others stop supporting others that are undesirable for her. Or, vice-versa, if everyone supports some undesirable candidate in exchange for others to help a candidate that they want to be selected. This type of manipulation involves a Pareto improvement, which is not surprising since we have already remarked that anonymous strategy-proof rules in that context are not efficient. Hence, if we change the equilibrium concept that we apply to the voting game in that example, and simply require that manipulations by agents must be a Nash equilibrium, then the mechanism is co-utile, at least in the sense that the equilibria of no other mechanism Pareto dominate those of the one we started with.

Moreover, notice that the type of improvement in question derives from a transaction between groups of agents who can gain from giving up some gains in exchange for some larger ones. And that in order to make this transaction a Nash equilibrium, the agents end up voting in a manner that does not allow to tell each agent from the others, and that none of the agents reveals their true identity. All of this is strongly reminiscent of the privacy preserving character of mechanisms that is sought after by co-utility.

Our general conclusion, then, is that knowledge of results in social choice has allowed us to better understand the ambitions of people working in co-utility, and that our still limited understanding of their goals has led us to re-evaluate some of our findings, and to understand that not all departures from the proposed rules by groups of agents must be treated as undesirable, if they end up approaching the collective results to efficiency. We hope that our attempt to build that bridge meets others, in the directions of interdisciplinary dialog.

Acknowledgements S. Barberà acknowledges financial support through grants ECO2014-53052-P and SGR2014-515, and Severo Ochoa Programme for Centres of Excellence in R&D (SEV-2015-0563). D. Berga acknowledges the support from grants ECO2016-76255-P, ECO2013-45395-R and 2014-SGR-1360. B. Moreno acknowledges financial support from grants and ECO2014-53767. D. Berga and B. Moreno thank the MOMA network under project ECO2014-57673-REDT.

References

1. Arribillaga, P., Massó, J., Neme, A. (2016) Not All Majority-based Social Choice Functions Are Obviously Strategy-proof. Mimeo UAB, pp. 1–48
2. Arrow, K.J.: Social Choice and Individual Values. Yale University Press, New Haven and London (1963)
3. Barberà, S.: Strategy-proof social choice. In: Arrow, K.J., Sen, A.K., Suzumura, K. (eds.) Handbook of Social Choice and Welfare, vol. 2, pp. 731–831. Elsevier, Amsterdam (2010)
4. Barberà, S., Berga, D., Moreno, B.: Individual versus group strategy-proofness: when do they coincide? J. Econ. Theory **145**, 1648–1674 (2010)
5. Barberà, S., Berga, D., Moreno, B.: Group strategy-proofness in private good economies. Am. Econ. Rev. **106**, 1073–1099 (2016)
6. Barberà, S., Berga, D., Moreno, B.: Immunity to credible deviations from the truth. Math. Soc. Sci. (2016). doi:10.1016/j.mathsocsci.2016.11.002
7. Barberà, S., Gul, F., Stacchetti, E.: Generalized median voter schemes and committees. J. Econ. Theory **61**, 262–289 (1993)
8. Barberà, S., Sonnenschein, H., Zhou, L.: Voting by committees. Econometrica **59**, 595–609 (1991)
9. Brandt, F., Conitzer,V., Endriss, U., Lang, J., Procaccia, A.D.: Introduction to computational social choice. In: Brandt, F., Conitzer, V., Endriss, U., Lang, J., Procaccia, A.D. (ed.) Handbook of Computational Social Choice, pp. 1–22. Cambridge University Press, Cambridge (2015)
10. Domingo-Ferrer, J., Martínez, S., Sánchez, D., Soria-Comas, J.: Co-utility: self-enforcing protocols for the mutual benefit of participants. Eng. Appl. Artif. Intell. **59**, 148–158 (2017)
11. Domingo-Ferrer, J., Sánchez, D., Soria-Comas, J.: Co-utility: self-enforcing collaborative protocols with mutual help. Prog. Artif. Intell. **5**(2), 105–110 (2016)
12. Farquharson, R.: Theory of Voting. Yale University Press, New Haven (1969)
13. Gibbard, A.: Manipulation of voting schemes: a general result. Econometrica **41**, 587–601 (1973)

14. Mc Lean, I., Urken, A.: Classics of Social Choice. The University of Michigan Press, Michigan (1995)
15. Li, S.: Obviously strategy-proof mechanisms (2017). doi:10.2139/ssrn.2560028
16. Moulin, H.: On strategy-proofness and single peakedness. Public Choice **35**, 437–455 (1980)
17. Nisan, N.: Introduction to mechanism design (for computer scientists). In: Nisan, N., Roughgarden, T., Tardos, E., Vazirani, V. (eds.) Algorithmic Game Theory. Cambridge University Press, Cambridge (2007)
18. Pattanaik, P.K.: Counter-threats and strategic manipulation under voting schemes. Rev. Econ. Stud. **43**(1), 11–18 (1976)
19. Pattanaik, P.K.: Threats, counter-threats, and strategic voting. Econometrica **44**(1), 91–103 (1976)
20. Peleg, B.: Almost all equilibria in dominant strategies are coalition-proof. Econ. Lett. **60**, 157–162 (1998)
21. Peleg, B., Sudhölter, P.: Single-peakedness and coalition-proofness. Rev. Econ. Design **4**, 381–387 (1999)
22. Satterthwaite, M.: Strategy-proofness and arrow's conditions: existence and correspondence theorems for voting procedures and social welfare functions. J. Econ. Theory **10**, 187–217 (1975)
23. Vickrey, W.: Utility, strategy and social decision rules. Q. J. Econ. **74**, 507–535 (1960)

Part II
Applications of Co-utility

Co-utile P2P Anonymous Keyword Search

Josep Domingo-Ferrer, Sergio Martínez, David Sánchez and Jordi Soria-Comas

Abstract Web search engines have become an essential tool to find online resources. However, their use can reveal very detailed information about the interests of their users. In this chapter, we detail co-utile protocols for exchanging queries between users in an anonymous way. Specifically, by exchanging queries with other users, the query profile associated to each user is masked and approximates the average group interests. However, the query exchange protocol has a cost. To thwart free-riders (who take advantage of the protocol to mask their profile but are reluctant to help others) we employ a co-utile reputation management system. The theoretical analysis is complemented with empirical results obtained from an implementation in a simulated multi-agent environment, which illustrates how co-utility can make cooperation self-enforcing and improve the agents' welfare.

1 Introduction

Web search engines (WSEs) have become an essential tool to find online resources. However, their use can reveal very detailed information about the interests (profiles) of their users. The users' defense consists in making their profile diverse enough so that the WSE cannot determine their interests. In essence, to mask their profiles, the users must hide the real queries they are interested in among a set of queries about diverse topics they are not interested in.

J. Domingo-Ferrer (✉) · S. Martínez · D. Sánchez · J. Soria-Comas
UNESCO Chair in Data Privacy, Department of Computer Science and Mathematics,
Universitat Rovira i Virgili, Av. Països Catalans 26, 43007 Tarragona,
Catalonia, Spain
e-mail: josep.domingo@urv.cat

S. Martínez
e-mail: sergio.martinezl@urv.cat

D. Sánchez
e-mail: david.sanchez@urv.cat

J. Soria-Comas
e-mail: jordi.soria@urv.cat

J. Domingo-Ferrer and D. Sánchez (eds.), *Co-utility*, Studies in Systems,
Decision and Control 110, DOI 10.1007/978-3-319-60234-9_4

The users can perform the hiding on their own, by generating random fake queries and submitting them along with their real queries [3, 9, 12, 15]. Such a standalone approach has several downsides: on the one hand, fake queries overload the WSE, so, if every agent followed this approach, the WSE performance would strongly degrade; on the other hand, it is not so easy to produce fake queries that are plausibly indistinguishable from real ones [10].

Relying on a set of peer agents (as proposed in [2, 11]) is an alternative that is free from the previous downsides. Consider several agents with similar privacy interests. To mask her profile, an agent can ask another agent to submit her query to the WSE rather than submitting it herself. Also, the agent can submit to the WSE real queries originated by other agents. In this way, a rational cooperation between peer agents emerges [4, 5] and all of them manage to blur their profile of interests without having to submit fake queries.

In this chapter, we analyze this cooperation under the co-utility framework [6, 7] (see also the first chapter of this book). We propose co-utile protocols for exchanging queries between users in an anonymous way. Specifically, by exchanging queries with other users, the query profile associated to each user is masked and approximates the average group interests. However, the query exchange protocol has a cost for the agents submitting the queries to the search engine. To thwart free-riders (who take advantage of the protocol to mask their profile but are reluctant to help others), we employ a co-utile reputation management system [8] (also described in the second chapter of this book), so that both the query submission protocol and the reputation management mechanism are self-enforcing and mutually beneficial for the involved peers.

The theoretical details of the protocols are complemented with empirical results obtained from an implementation in a simulated multi-agent environment, which illustrates how co-utility can make cooperation self-enforcing and improve the agents' welfare in this specific scenario.

2 Single-Hop Query Submission Game with Two Agents

Within the anonymous keyword search scenario, the primary utility agents want is functionality: they want to get their queries answered. A secondary utility is privacy: they may also wish their interest profiles to stay private w.r.t. the WSE. Note that any WSE can be expected to try to profile its users (typically for marketing purposes) according to the topics mentioned in the queries they perform. We will call atomic utilities the two components of an agent's utility, namely functionality and privacy. These two atomic utilities are not easily reconciled, because an agent alone cannot optimize both of them simultaneously.

The first scenario/game we analyze, which we call *single-hop query submission*, is restricted to two agents: the initiator of the query (I) and the responder to a query submission request (R). The initiator I is interested in functionality (having her query answered) and privacy w.r.t. the web search engine (keeping her interests private).

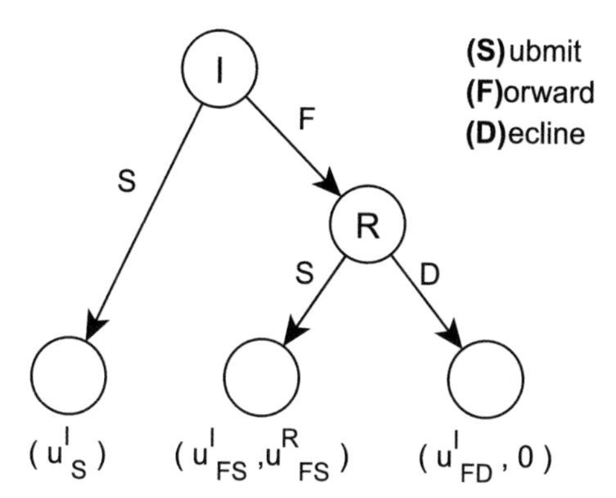

Fig. 1 Tree showing the possible interactions in the single-hop anonymous query submission game between two agents: initiator (I) and responder (R). To keep the figure compact, we only represent the payoffs of the agents active in each interaction: for interaction (S), the payoff of any agent other than the initiator is 0; for the other interactions, only the payoffs of the initiator and the active responder are represented, the rest of agents' payoffs being 0

If the new query diversifies her interest profile (for example, because she had never asked a similar query), I is likely to submit it herself to the WSE; otherwise, I will ask R to submit the query on her behalf. The responder R is only concerned about his privacy (whether I's query is useful to blur R's profile w.r.t. the WSE).

The interaction among I, R and the WSE is as follows:

- I chooses between submitting the query to the WSE herself or forwarding the query to R for submission.
- Upon query reception, R chooses between submitting the query to the WSE (and returning the results to I) or declining to submit it.

These interactions are illustrated in Fig. 1. Since (S) and (F, D) do not have any effect on the privacy or the utility of the responder, we assume that the responder's payoff in these protocols (u^R_S and u^R_{FD}, respectively) is 0. It is also reasonable to assume that $u^I_{FS} > u^I_{FD}$, as both (F, S) and (F, D) are equivalent in the initiator's eyes regarding privacy w.r.t. the WSE, but (F, S) is better in terms of functionality.

Given the above assumptions about the payoffs of the agents under different protocols, we can determine the possible co-utile protocols in the game tree shown in Fig. 1. Since the payoff of the responder when he does not participate in the protocol (that is, when the initiator submits the query to the WSE herself) is 0, (F, D) is not a co-utile protocol because the responder's payoff is the same as in (S). The only possible co-utile protocols are:

- (F, S), which is co-utile if the responder's payoff is strictly greater than in (S), that is $u^R_{FS} > u^R_S = 0$, and (F, S) is self-enforcing, which occurs if $u^R_{FS} > u^R_{FD}$ (the responder improves her privacy by submitting the query, which happens only when the query diversifies R's query profile) and $u^I_{FS} > u^I_S$ (directly submitting the query decreases the initiator's privacy). Notice that, under the previous conditions, (F, S) is also Pareto-optimal.

- (S), which is co-utile (actually just utile) when any of the following conditions is satisfied: $u_S^I > \max\{u_{FS}^I, u_{FD}^I\} = u_{FS}^I$ (by directly submitting her query, the initiator improves her privacy); or $u_S^I = u_{FS}^I$ and $u_{FS}^R = 0$ (the initiator does not decrease her privacy by directly submitting her query and the responder does not increase his privacy by submitting the initiator's query).

The following proposition summarizes the previous discussion.

Proposition 4.1 *Protocol* (F, S) *is co-utile when* $u_{FS}^R > 0$ *and* $u_{FS}^R > u_{FD}^R$ *and* $u_{FS}^I > u_S^I$. *In fact, under these conditions* (F, S) *is strictly co-utile.*

While we have been able to obtain Proposition 4.1 without specifying the form of the agents' utility functions, we need to do so in order to compute the agents' payoffs.

The initiator is interested in getting her query answered by the WSE and keeping her interests private w.r.t. the WSE. Thus, the initiator's utility under a protocol P can be expressed as $u_I(P) = u_I(f_I(P), p_I(P))$, where $p_I(P)$ quantifies the effect of the query submission on the initiator's privacy and $f_I(P)$ is an indicator reflecting whether protocol P gives the expected functionality to the initiator:

$$f_I(P) = \begin{cases} 0 & \text{agent } I \text{ does not get the expected functionality;} \\ 1 & \text{otherwise.} \end{cases}$$

As to the responder, his only interest is his privacy w.r.t. the WSE; therefore, his utility can be written as $u_R(P) = u_R(p_R(P))$, where $p_R(P)$ quantifies the effect of the query submission on the responder's privacy.

To measure the privacy of an agent w.r.t. the WSE, we look at the agent's query profile, Y, which is the distribution of topics of the queries submitted so far by the agent. The query profile characterizes the exposure level (that is, privacy loss) of the agent's interests towards the WSE: if an agent's profile just contains one or a few dominant topics, her/his preferences are obvious; on the other hand, if it contains a variety of topics all appearing with the same frequency, the agent's preferences remain uncertain. We measure the agent's privacy level by computing the Shannon's entropy (that is, uncertainty) of her/his profile. Thus, the effect of a query q can be measured as the variation of the profile entropy caused by q:

$$p_I(P) = H(Y_I \cup \{q\}) - H(Y_I);$$

$$p_R(P) = H(Y_R \cup \{q\}) - H(Y_R).$$

Even if the Shannon's entropy of an empty profile is not defined mathematically, by convention we set it to $\log_2 \tau$, which is the maximum value of the entropy of a distribution over τ possible topics; the reason is that an empty profile discloses no information about the interests of the agent. For non-empty profiles, the maximum entropy/privacy $\log_2 \tau$ is reached if all the topics in the profile appear with the same frequency; the rationale is that such a flat profile does not disclose any particular preferences by the agent.

3 Single-Hop Query Submission Game with Multiple Responders

So far, we have considered only two agents (initiator and responder). While this simplification is useful to illustrate how the anonymous query submission game works, it also strongly limits cooperation and therefore the privacy gain of the agents: the two agents will cooperate to submit each other's queries only if their interests are complementary. To overcome this limitation, more agents are needed. Here we assume that there is one initiator (I) and n responders ($R_1, \ldots, R_n$).

The interactions and utilities of the agents are the same as in the previous case; the difference is that, when forwarding the query to another agent, the initiator has to select one among the n responders. Figure 2 illustrates the possible interactions in the scenario with multiple responders. Notice that action F_i has no effect on the privacy or the functionality of responder R_j, with $j \neq i$. Thus, it is reasonable to assume that the utility of R_j, with $j \neq i$, is zero under F_i. Also, (S) has no effect over any of the responders, so that the utility of R_i under (S) should be zero for all i. Finally, when responder R_i declines the submission, the effect on the privacy and the functionality of all responders should also be zero.

If we analyze the conditions under which co-utile protocols exist in this game, we have:

Proposition 4.2 *Protocol* (F_i, S) *is co-utile when* $u^{R_i}_{F_i S} > 0$ *and* $u^{R_i}_{F_i S} > u^{R_i}_{F_i D}$ *and* $u^{I}_{F_i S} > u^{I}_{S}$.

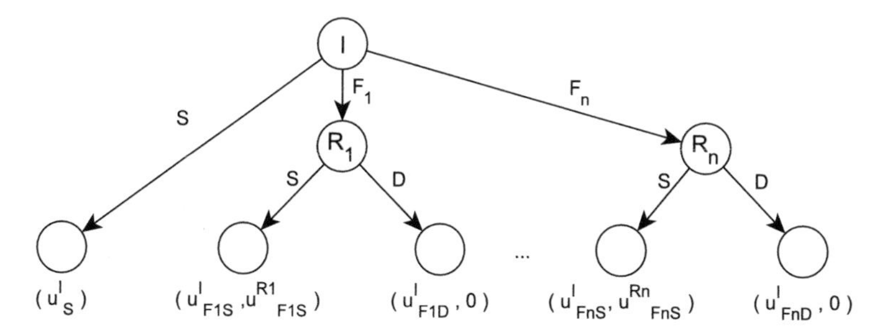

Fig. 2 Tree showing the possible interactions in the single-hop anonymous query submission game with n responders. To keep the figure compact, we only represent the payoffs of the agents active in each interaction

Unlike in the game with only two agents, none of the protocols (F_i, S), for $i = 1, \ldots, n$, is strictly co-utile because each responder maximizes his utility at a different terminal node.

Since the initiator is unaware of the state of the other agents, she cannot anticipate whether a specific responder will submit or decline her query. In a game with incomplete information as this one, the initiator must check her prior knowledge on the other agents and choose the responder that she expects to be most likely to submit her query. However, following a protocol that is expected to be co-utile is not a satisfactory solution if the expectations are not fulfilled and the query is declined (in this case the protocol actually turns out to be not co-utile).

Also, it does not make sense for the initiator to stop when the query submission is declined by the selected responder. After a refusal, the initiator updates her knowledge about the other agents (by excluding the declining responder) and tries again. This trial and error process has a drawback: it can defer the query submission too long. To prevent this, the initiator should include time among the factors that shape utility, more precisely as part of the functionality atomic utility: the initiator does not only want her query answered, she wants to get the answer within a certain time interval. To account for time, we assume that the response from the WSE is immediate when the initiator submits the query herself, while forwarding it to another agent for submission and getting the answer back implies some delay.

Each new trial implies a different game, in which the utilities are updated with respect to the previous trial game: not only the declining responder in the previous game is excluded, but the initiator's utility is decreased to reflect the passing of time. To prevent the query submission from being deferred too long, there must be a time t such that, regardless of privacy, getting the query answered at time t is always preferable to getting it answered at time $t + 1$. Thus, at time t, (S) is preferred by the initiator to (F_i, S) for any i.

Figure 3 depicts the initial game tree for a scenario with an initiator and two responders, where the initiator includes time as part of functionality. Figure 4 shows the updated game tree assuming that the initiator has played F_1 and responder R_1

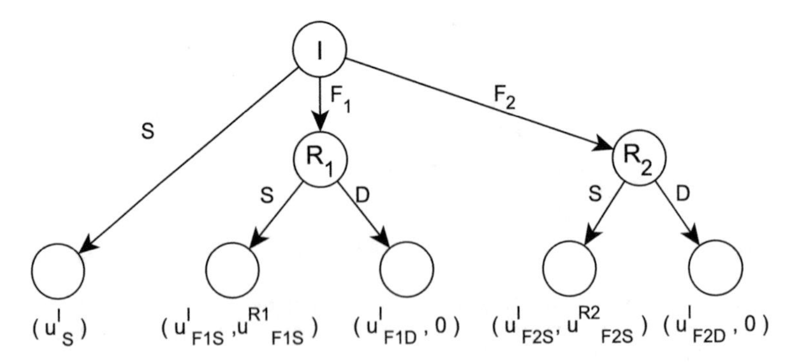

Fig. 3 Tree of interactions in the case of an initiator and two responders at time $t = 0$. To keep the figure compact, we only represent the payoffs of the agents active in each interaction

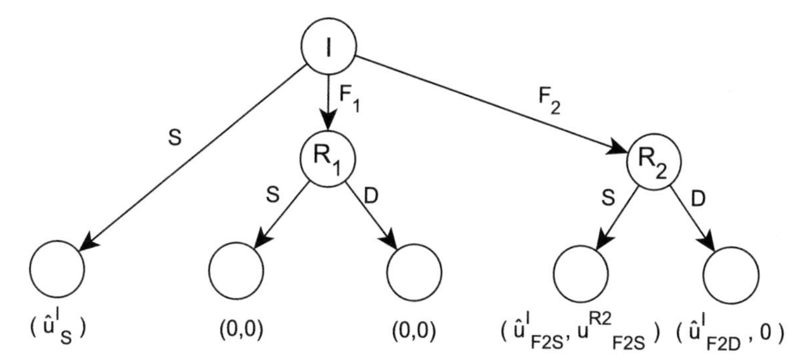

Fig. 4 Updated tree of interactions in the case of an initiator and two responders at time $t = 1$, assuming that responder R_1 declined to submit the initiator's query. To keep the figure compact, we only represent the payoffs of the agents active in each interaction. Due to the elapsed time, it holds that $\hat{u}^I_S \leq u^I_S$, $\hat{u}^I_{F_2S} \leq u^I_{F_2S}$ and $\hat{u}^I_{F_2D} \leq u^I_{F_2D}$, where u^I_S, $u^I_{F_2S}$ and $u^I_{F_2D}$ are the utilities in Fig. 3

has declined the submission. This is reflected in the initiator's payoffs, which are reduced by one due to the passed time: the utilities $\hat{u}^I$ in Fig. 4 are smaller than the utilities u^I in Fig. 3.

Let us now examine how functionality can account for passing time. Clearly, functionality can no longer be a binary function in $\{0, 1\}$. Let us assume it takes values in $[0, 1]$, so that the actual value measures the initiator's satisfaction at obtaining the query answer at time t:

$$f_I(P; t) = \begin{cases} 0 & \text{the initiator does not get the query answered at time } t; \\ \alpha_t \in [0, 1] & \text{satisfaction level when getting the answer at time } t. \end{cases}$$

A stricter way to define $f_I(P; t)$ in order to prevent too long a deferral of query submission is to set a timeout, T that, if reached, makes (S) the best choice for the initiator; that is, the query answer is only useful if it comes no later than T. To model this behavior, we can set the functionality to $-\infty$ when time T is reached; furthermore, to abstract from the initiator's beliefs, we can consider the functionality to be constant across time until T and to become $-\infty$ thereafter:

$$f_I(P; t) = \begin{cases} 0 & \text{if } t < T \text{ and the initiator does not get the query answered;} \\ 1 & \text{if } t < T \text{ and the initiator gets the query answered;} \\ -\infty & \text{if } t > T. \end{cases}$$

With the above functionality, the initiator's utility does not change for times until T, so that the reasonable way to act for her is to keep trying different responders until either the query is submitted to the WSE or time T is reached. If T is reached, then the initiator submits the query herself. Although Expression (3) is reasonable for the initiator's functionality atomic utility, the actual expression depends on the initiator's preferences: if the functionality gradually decreases with time, the beliefs about the other agents' preferences become more relevant, as it may be better for the initiator to submit the query herself, even before T, if she believes the other agents are likely to decline.

4 Multi-hop Query Submission Game

The single-hop query submission protocol discussed above is effective at protecting the privacy of agents against the WSE. However, since the initiator directly forwards her query to the responders, she also discloses her interests to those responders. Thus, if agents are not only interested in protecting their privacy toward the WSE, but also against their peers, the single-hop protocol is not co-utile anymore. Actually, collaborating with peer agents to avoid being profiled by the WSE can hardly be satisfactory if it has the side effect of being profiled by one's peers.

To realize co-utility when agents want privacy against the WSE and the other agents, a more complex game needs to be designed whereby the identity of the query initiator is not only hidden to the WSE, but also to all other agents. We can achieve this by extending the actions available in the single-hop game as follows: an agent that receives a query to be submitted may now submit it, decline it or *forward it to another peer*. Specifically, forwarding the query to another agent is an alternative to declining it in case the responder does not wish to submit the query to the WSE (because doing so does not increase his privacy). The result of allowing query forwarding is that a responder agent who receives a query no longer knows whether he is receiving it from the initiator or from another responder. Figure 5 illustrates the tree of interactions of this *multi-hop* game in the case of three agents.

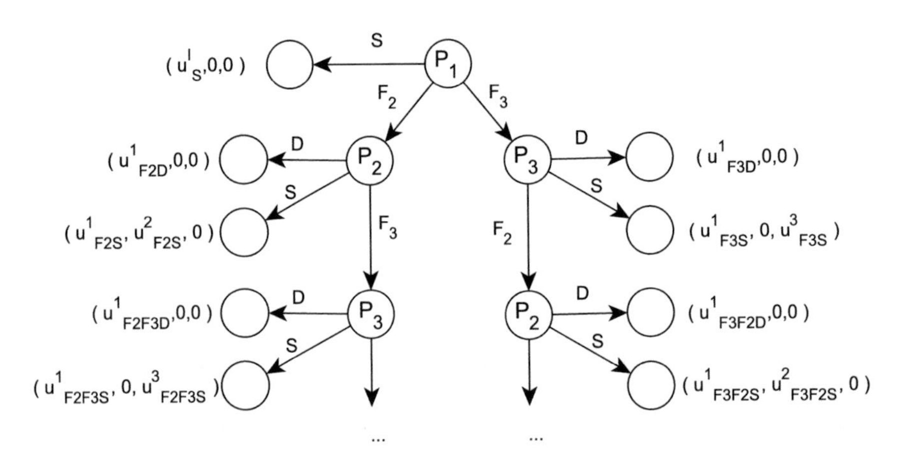

Fig. 5 Tree showing the possible interactions of the multi-hop query submission game with three agents

If, in the single-hop game with multiple responders, it could take too long for a query to be answered (in case many responders needed to be tried), in the multi-hop game the situation can be even worse. If agents keep forwarding the query without submitting it, the submission can be deferred indefinitely. Like in the previous game, including time in the initiator's utility is a way to deal with this potential problem. Thus, similarly as we did in the single-hop game, to abstract from the initiator's beliefs about the other agents, we assume that the initiator is indifferent to the query response time as long as the answer is received before T; when time T is reached, her utility becomes $-\infty$, which makes direct submission by the initiator the most preferable protocol.

On the other hand, in order to remain perfectly private w.r.t. other peers in the network (or any external observer), agents will also be interested to add uncertainty to the profile that results from the whole set of queries they send (those forwarded to other agents or submitted to the WSE). Note that this assumes the worst case in which a network adversary is capable of collecting all queries initiated, forwarded or submitted by each agent. We denote the profile that results from accumulating these queries by Z and, like for Y, we use Shannon's entropy to measure how disclosive it is. By incorporating this additional privacy concern into the privacy atomic utility, we have the following updated expressions that also consider the variation of the entropy that q produces in Z when q is submitted to the WSE or forwarded to another agent:

$$p'_I(P) = p_I(P) + H(Z_I \cup \{q\}) - H(Z_I);$$

$$p'_R(P) = p_R(P) + H(Z_R \cup \{q\}) - H(Z_R).$$

From a co-utility perspective, the possibility of forwarding the query to another agent masks the initiator of the query, thereby preserving her privacy versus the other agents. On the other hand, the privacy concerns of agents w.r.t. external observers make forwarding preferable to declining to any responder agent: forwarding agents also increase the privacy of their external profiles (Z) because the forwarded queries add uncertainty to the queries they really originate. Thus, in a multi-hop protocol with several forwarding hops ending in a submission, not only the initiator and the submitter take their best possible actions, but the forwarders also do. In this sense, the protocol is strictly co-utile (if we consider only the agents participating in it).

5 On the Availability of Agents with Appropriate Types

The above co-utile protocols are viable provided that agents with appropriate preferences/types are available. In this section, we discuss how these types are, at the end, the most common ones in a real scenario. To do so, we analyze the system in the most extreme cases.

The first extreme case happens when all agents have empty profiles towards the WSE and/or towards all the other agents because they have not (directly) submitted any query so far. We call this the "cold start" case. As discussed in Sect. 2, an empty profile offers perfect privacy (to which we assigned by convention maximum entropy in Sect. 2) because it does not disclose any preference of the agent. According to this, one may assume that agents will always prefer to rely on others to submit their queries and will decline submitting queries of others because this would decrease the entropy of their profiles towards the WSE. This "free-riding" attitude results in all requests for query submission being declined. However, this deadlock cannot be sustained very long: if the initiator has a query that does not make anyone else more private (and, thus, no one wants to forward it, let alone submit it) and if the response time is part of the initiator's utility, she will eventually be forced to submit the query herself. In this way, the initiator's profile stops being empty and its entropy falls sharply (the agent's query is completely exposed); thus, the initiator becomes highly motivated to submit queries from other agents in order to flatten her profile and thereby increase its entropy.

A similar analysis and conclusion apply to the other extreme case that we dub "happy laziness". This situation happens when most of the agents have a non-empty flat profile and, because of this, they prefer not to submit or forward queries from others. "Happy laziness" however comes to an end as soon as some agents initiate new queries: since they cannot find anyone willing to forward or submit them, they are forced to submit them themselves. As in the "cold start" case, this action unflattens their profiles and makes them willing to collaborate with others to increase again the entropy of their profiles.

6 Anonymous Query Submission Considering Agents' Reputation

In the protocols above, the utility of the initiating agent $\mathscr{P}_i$ (i.e., the one who wants to have a query q answered), can be defined as the aggregation of the functionality and privacy atomic utilities, which are a function of the query q:

$$u_i(q, f_i(\cdot), Y_i, \alpha_i^H) = f_i(q) + \alpha_i^H \cdot \Delta(H(Y_i), q). \quad (1)$$

The first term, $f_i(q)$, represents the functionality that the agent gains from having q answered, which is 1 if the query is answered and 0 otherwise. The second term quantifies the privacy gain or loss w.r.t. the WSE as a result of the submission of the query to the WSE (by $\mathscr{P}_i$ herself or by another agent). Specifically, Y_i is $\mathscr{P}_i$'s profile of submitted queries (which is what the WSE sees on $\mathscr{P}_i$); it is represented as a distribution of topics extracted from the queries submitted so far by agent $\mathscr{P}_i$. From $\mathscr{P}_i$'s perspective, Y_i characterizes the exposure level (i.e., privacy loss) of the agent's interests towards the WSE. The function H is Shannon's entropy and, thus, $H(Y_i)$ measures the uncertainty (i.e., privacy) of the profile. Note that the highest privacy level is reached when the profile is empty, because no information about the profile has been disclosed. For non-empty profiles, the maximum entropy/privacy is reached if all topics appear with the same frequency, because the agent's interests remain unspecific. The $\Delta(H(Y_i), q)$ function represents the variation (positive or negative) of the agent's privacy in terms of the profile's entropy $H(Y_i)$ as a result of $\mathscr{P}_i$ submitting herself the query q to the WSE; if the query is finally submitted by any other agent, $\Delta(H(Y_i), q) = 0$, because the WSE will not learn anything new about the initiator.

Finally, the α_i^H coefficient normalizes the scales of functionality and privacy (so that these heterogeneous features can be coherently compared) and also weights their relative importance to agent $\mathscr{P}_i$. By properly configuring this coefficient, we can make functionality dominate (which is a natural choice).

For the responder $\mathscr{P}_j$, the only utility that matters is his own privacy w.r.t. the WSE and how it is modified by submitting the query q of the initiator $\mathscr{P}_i$. Thus, to derive the utility for $\mathscr{P}_j$, we set $f_j(q) = 0$ in Expression (1) and we obtain the following simplified expression:

$$u_j(q, Y_j) = \Delta(H(Y_j), q). \tag{2}$$

Co-utility in the scenario depicted above may be hampered by additional hindrances that may arise in a real setting. In particular, submitting a query q may result in a negative cost (e.g., spent bandwidth or money) to the agents. In this case, the utility functions for the initiator and responder agents need to incorporate this cost $o(q)$ as an additional negative atomic utility:

$$u_i(q, f_i(\cdot), o_i(\cdot), Y_i, \alpha_i^H, \alpha_i^o) = f_i(q) + \alpha_i^H \cdot \Delta(H(Y_i), q) - \alpha_i^o \cdot o_i(q); \tag{3}$$

$$u_j(q, o_j(\cdot), Y_j, \alpha_j^o) = \Delta(H(Y_j), q) - \alpha_j^o \cdot o_j(q). \tag{4}$$

For the initiator agent, the cost $o_i(q)$ constitutes an additional incentive to prefer forwarding the query (which results in $o_i(q) = 0$) rather than submitting it herself. For the responder agent, if the privacy gain $\Delta(H(Y_j), q)$ brought to him by the query q does not dominate (considering the weighting coefficient) the cost $o_j(q)$ of the query, the outcome of the utility function is negative and submitting the query "for free" is not rational for him anymore. Thus, in this case, the protocol is no longer self-enforcing, let alone co-utile.

In order to neutralize this negative payoff, agent reputations can be used to ensure collaboration in a way that the costs associated to query submission are fairly distributed throughout the network.

Specifically, we propose to compute the reputation t_i of an agent $\mathscr{P}_i$ as the difference between the cost incurred by $\mathscr{P}_i$ when submitting queries on behalf of other agents and the cost incurred by other agents as a result of submitting $\mathscr{P}_i$'s queries. Assuming the cost of a query submission is uniform for all agents (which implies that agents calculate and evaluate reputations in a similar way), if an agent $\mathscr{P}_i$ has a reputation t_i greater than 0, it means that she has helped others more than what others have helped her.

To evaluate whether the collaboration record of an initiator agent $\mathscr{P}_i$ compensates the cost of helping her, a responder agent $\mathscr{P}_j$ can take into account the reputation t_i of the initiator agent. In our case study, this yields:

$$u_j(q, o_j(\cdot), Y_j, t_i, \alpha_j^o, \alpha_j^t) = \Delta(H(Y_j), q) - \alpha_j^o \cdot (o_j(q) - \alpha_j^t \cdot t_i). \quad (5)$$

According to this expression and leaving aside the privacy gain, $\mathscr{P}_j$ will get a positive utility outcome if the reputation t_i of the initiator compensates the negative cost of the query $o_j(q)$. Likewise, $\mathscr{P}_j$ will decide to refuse submitting queries from any agent $\mathscr{P}_i$ whose reputation (multiplied by the corresponding weight) is less than the relative cost of the query $o_j(q)$. Since the cost is an unbounded value and the reputation is a normalized value, the α_j^t coefficient is necessary to make both values comparable.

Note, however, that the relative importance given to privacy with respect to the cost/reputation difference (weighted by the α_j^o coefficient) may even compensate a negative difference, because an agent may still help a peer with low reputation if the former attaches a lot of importance to her own privacy.

Reputations in [8] are biased towards positive values; that is, in general, initiator agents only get help from other peers if they have already helped other agents (i.e., submitted their queries) more than what others have helped them. In this way, non-collaborative agents become motivated to help others in order to increase their own reputation beyond the relative cost of their queries; otherwise they will be forced to submit their queries themselves, thereby incurring two negative payoffs: query submission cost and loss of privacy. Since the sum of these negative payoffs will severely decrease the agent's utility, helping others becomes the rational choice (helping others certainly has a cost, but it also increases both the reputation and the privacy of the helper).

Thanks to the compensation of costs via reputation, the above reputation-based query submission protocol is strictly co-utile in spite of the query costs. To see this, note that, if the responder $\mathscr{P}_j$ agrees to submit, it is because this is the action among submitting or refusing the initiator's query that brings the maximum utility to $\mathscr{P}_j$. In this case, the outcome for the initiator $\mathscr{P}_i$ is also maximum (because she avoids the privacy loss and the cost of submitting her query).

Ideally, to balance the costs and the social welfare of all agents, these should achieve and maintain a bounded reputation t_{target} slightly greater than 0 that is sufficient to get help from other peers. In this respect, as discussed in [8], agents can view their own reputations as atomic utilities: they should measure the reputation gain that each action of the protocol brings them and how far their actual reputation is from the target reputation t_{target} they want to achieve. Thus, the aggregated utility functions are as follows:

$$\begin{aligned} &u_i(q, f_i(\cdot), o_i(\cdot), Y_i, \alpha_i^H, \alpha_i^o, \alpha_i^{t'}) = \\ &= f_i(q) + \alpha_i^H \cdot \Delta(H(Y_i), q) - \alpha_i^o \cdot o_i(q) + \alpha_i^{t'} \min((t_{i,1} - t_{i,0}), (t_{target} - t_{i,0})); \end{aligned} \quad (6)$$

$$u_j(q, o_j(\cdot), Y_j, \alpha_j^o, \alpha_j^t, \alpha_j^{t'}) =$$
$$= \Delta(H(Y_j), q) - \alpha_j^o \cdot (o_j(q) - \alpha_j^t \cdot t_i) + \alpha_j^{t'} \cdot \min((t_{j,2} - t_{j,1}), (t_{target} - t_{j,1})), \quad (7)$$

where α_*^H, α_*^o are the weights attached by agents to privacy and cost, respectively, α_*^t is the weight attached to the reputation of other agents, $\alpha_*^{t'}$ is the weight attached to one's own reputation, $t_{i,0}$ is $\mathscr{P}_i$'s reputation before starting the protocol, $t_{i,1}$ is $\mathscr{P}_i$'s reputation after requesting and obtaining $\mathscr{P}_j$ to submit q, $t_{j,1}$ is $\mathscr{P}_j$'s reputation before accepting to submit q on $\mathscr{P}_i$'s behalf and $t_{j,2}$ is $\mathscr{P}_j$'s reputation after submitting q. Note that $\mathscr{P}_i$ and $\mathscr{P}_j$ can accurately anticipate $t_{i,1}$ and $t_{j,2}$, respectively (they do not need to use estimates); this is because in our case study reputation is objective, based on help provided and received.

Considering reputations as atomic utilities as in Expressions (6) and (7) further reinforces the co-utility of the protocol, because now responder agents are additionally motivated to collaborate (in order to increase their own reputations) and initiator agents need to make sure they have high enough reputations to afford their queries to be submitted by other peers (which decreases their reputation). Moreover, as discussed in [8], viewing reputations as utilities also contributes to a fair allocation of query submission costs among the peers.

6.1 Other Benefits of Co-utile Reputation Enforcement

Co-utile protocols are viable given the availability of agents with appropriate types/preferences. In this section we illustrate in our case study how the use of reputation makes the protocol realistically self-enforcing by widening the range of agent types and situations in which the query submission protocol remains co-utile.

First, let us consider the case in which a new agent with an empty profile (i.e., she has not submitted any query so far to the WSE) joins the network. Recall that an empty profile has maximum privacy because it does not disclose any interest of the agent. Even neglecting the cost associated to the query submission, such a newcomer will always prefer to rely on others to submit her queries and will refuse submitting queries from others because doing so will decrease her privacy. In plain words, the newcomer becomes a rational "free rider".

This is an endemic problem of many P2P co-utile protocols [8, 13] that can be fixed using reputation. With the reputation mechanism, newcomers have zero global reputation, because no one has interacted with them and the distributed reputation calculation is based on the local reputations, which are zero for newcomers. Thus, in the reputation-based protocol discussed in this case study, responder agents will systematically refuse requests for query submission from free riders. Given that the functionality (i.e., to have the query answered) is likely to dominate the other utilities, the free-riding agent has only two options left:

- She submits her query herself. This action, in turn, will create a non-empty profile for the agent and a sharp decline of the profile's privacy because the agent is now completely exposed to the WSE. As a result, the agent will become highly motivated to contribute to the system by submitting other agents' queries in order to flatten her profile again.
- She accepts a query request from another agent. This results both in a cost and in a decrease of her privacy, even though her profile gives a distorted view of her interests towards the WSE because the query is unlikely to match her true interests. However, this action will also increase the agent's reputation, who will now be able to find another peer willing to submit her query.

In both cases, newcomers with empty profiles become collaborative members of the network in a self-enforcing way. Note that newcomers with non-empty profiles (i.e., who have already submitted queries before joining the network) are not problematic at all because they are already motivated to accept queries from other peers to flatten their profiles.

The above discussion can be extended to an extreme scenario in which all agents join the system at once with empty profiles and zero global reputations. One might think that this setting brings the system to a standstill in which any request for query submission by any agent is refused and agents do not want to submit their own queries. However, it is easy to see that, thanks to functionality dominating the other utilities, agents will end up submitting their own queries; in turn, this will instantly motivate them to submit queries from peers, which will also increase their reputations and the chances that other agents accept their requests.

A different scenario which might also seem standstill-prone occurs when agents achieve a non-empty but flat profile (and probably a high reputation) thanks to the number of query requests they have satisfied; at this point, they might be tempted to stop collaborating. In the unlikely case that all agents reach such a state, they will be unable to find peers willing to submit their queries, regardless of the reputations they have; as a result, they will be forced to submit their own new queries themselves, which in turn will unflatten their profiles and make them again willing to collaborate. If only some of the agents reach this "too happy" state, their reputation will decrease if they ask other peers to submit their new queries; when their reputation becomes too low, they will be unable to find any other peer willing to submit their queries, which will force them to submit them themselves; this will make them again willing to collaborate.

Finally, the fact that agents may also consider their own reputations as atomic utilities has two virtuous effects. On the one hand, since the responder increases his reputation as a result of the submission whereas the initiator decreases her reputation by the same magnitude, the reputation management guides the system towards a state in which most of the agents have a similar –target– reputation, which corresponds to a *fair* allocation of costs/negative payoffs. On the other hand, in the basic protocol, initiator agents randomly chose potential responders, who may or may not fulfill their requests according to their –secret– privacy needs (profile). With the reputation mechanism, initiator agents may also take into account the reputation of the potential responders so that, instead of choosing them randomly, they may choose agents with zero or low reputation, because (for an equal secret profile) these are more motivated to help in order to increase their own reputations (and thus be able to get help from others). In this way, the initiator increases the chances of finding a helpful responder.

7 Implementation and Experiments

In this section, we empirically study the behavior of the privacy-preserving query submission protocol and the influence of the reputation mechanism to foster collaboration between agents.

The agent profiles w.r.t. the WSE have been characterized by following the state of the art in user profiling [16]: profiles are represented as vectors of normalized weights, each one quantifying the relative interest of the agent in a certain topic. Each topic weight represents the aggregation of the semantics of the user's queries that fall into the topic, which is quantified by means of linguistic and semantic analyses (see details in [16]).

The simulations have been carried out with 900 agents. Each one has a set of queries for which she wishes to get an answer from the WSE. These queries were randomly picked from the query logs of real users available in the AOL data set [1], which consists of logs of the actual queries performed by users of the AOL's WSE during 3 months in 2006. In total, we compiled 20,914 individual queries to be performed by the agents through the life cycle of the system. In our simulations, agents iteratively select a query from their query logs and decide whether or not to follow the anonymous submission protocol to retrieve the result from the WSE according to their states (current profiles of submitted queries) and types (utility functions). When applicable, global reputations are computed after each query submission.

We have set up the initial profile of each agent to follow the distribution of topics defined by the query log associated to each agent; we do this so that the queries initially submitted by each agent reflect her inherent interests. With this configuration, the average entropy of the initial profile of the agents is 2.15.

We evaluated the behavior of the system in the following scenarios:

- *S1*: In the simplest scenario, neither costs nor reputations are considered. Agents only care about the functionality of the query and their own privacy. Their utility functions are those of Expressions (1) and (2). Utility coefficients have been set to make functionality dominate, so that queries will be *always* submitted, either by the initiators or by other agents, regardless of the privacy gain/loss.
- *S2*: In the second scenario, the cost of submitting a query is considered. The utility functions of agents follow Expressions (3) and (4). The cost of a query $o(q)$ for the submitting agent is quantified as $+1$. We performed several executions by varying the relative importance (weighting coefficient α^o) that agents attach to the submission cost in their utility functions.

- *S3*: In the third scenario, both costs and agent reputations (measured as the difference between costs borne and caused) have been considered. In this case, responder agents have the utility of Expression (5). Given that reputations are normalized in the [0..1] range and reputation values of all peers add to 1, to properly aggregate costs and reputations, the latter have been de-normalized by setting α_j^t = number_of_peers = 900. Again, we performed several simulations by modifying the relative importance of the cost (cost/reputation difference for responder agents).
- *S4*: In the fourth scenario, agents also consider how their own reputation would be affected by the actions they can take (i.e., query forwarding or submission). Initiator agents follow Expression (6) and responder agents follow Expression (7). The target reputation value has been set to the average reputation (t_{target} = 1/number_of_peers = 1/900), so that reputations will tend to be balanced and, thus, costs will be fairly allocated. To properly aggregate costs and reputations, we have set the coefficient $\alpha_*^{t'}$ = number_of_peers $\cdot\, \alpha_*^o$ for each simulation. In this manner, the difference of reputations is de-normalized (so that it can be aggregated with cost values) and given the same weight as agent reputations in scenario *S3*.

The system behavior has been evaluated according to the following aspects:

- Number of queries that have been submitted by following the co-utile protocol. To put this value in context, we also compute the ratio between this number of queries and the total number of queries that initiator agents decide to forward according to their types, that is, the number of queries that potentially can be submitted in a co-utile way. A rate around 1 would be desirable, since it means that the protocol remains self-enforcing for most situations and agent types.
- The standard deviation of agents' reputations. Since our reputation measures the difference between the cost incurred by an agent when submitting other agents' queries and the cost caused to other agents who agree to submit the agent's own queries, a low deviation from the average reputation (which by definition is 1/number_of_peers) indicates a fair allocation of costs resulting from agent collaboration across the network.
- The privacy of agents' profiles, measured as the average entropy of their profiles at the end of each simulation.

For Scenario *S1*, Fig. 6 shows the evolution (on average over all the agents) of the agents' privacy (Y-axis), as a function of the number of queries that have been submitted so far (X-axis). The horizontal line at the top shows the privacy upper bound (i.e., the maximum entropy of the profiles when all agents achieve a perfectly flat distribution of the 15 topic weights considered in their profiles, which is $\log_2 15 = 3.9$).

We can see that the agent profiles become more private w.r.t. the WSE at each new iteration of the protocol. The entropy $H(Y)$ grows logarithmically and becomes quite flat after 10,000 queries. It is important to note that, in practice, the limited variability of the queries (whose topics are not evenly distributed among the 15

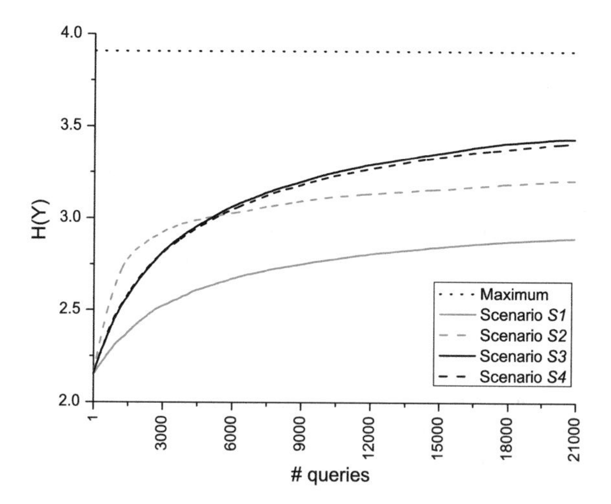

Fig. 6 Evolution of privacy (agents' profile entropy) for all scenarios with a cost coefficient of 0.25 (for *S2*, *S3*, *S4*)

Table 1 Scenario *S2*: Number and proportion of queries submitted to the WSE following the co-utile protocol, and average profile entropy (privacy) at the end of the simulation, for different cost coefficients

Cost coefficient	Co-utile queries (proportion)	Average profile entropy
4	0 (0)	2.15
1	398 (0.019)	2.62
0.25	2,930 (0.14)	3.20
0.063	8,788 (0.42)	3.47
0.016	15,816 (0.83)	3.35
0.004	14,341 (0.93)	3.05
0.001	11,910 (0.93)	2.95

possible topics) makes it difficult to achieve this theoretical upper bound (in which all agent profiles are perfectly flat). The average entropy of the agents' profiles at the end of the simulation is 2.9.

In this scenario, initiator agents preferred to forward 11,544 queries (from the total 20,914 queries), because submitting them directly would have decreased their privacy. From these, 10,806 queries (93.61%) were submitted by following the co-utile protocol; that is, in less than 7% of the cases, no other agent was willing to satisfy the initiator's request and the initiator was forced to submit the query herself (because the functionality utility dominates). This illustrates how, in absence of costs, the game is naturally co-utile for most agent types.

For Scenario *S2*, Table 1 shows the simulation results obtained by varying the relative importance given to the cost of performing a query to the WSE (α_i^o coefficient in Expression (3) for initiators and α_j^o coefficient in Expression (4) for responders).

We can clearly see how the more importance the agents attach to the cost of the queries, the more they tend to refuse submitting queries from other peers (even if refusal implies missing the chance of a privacy gain). As a result, initiators are forced to ask more peers and, if none of the latter is willing to satisfy their requests, initiators will end up submitting their queries themselves. This makes the system not co-utile for many agent types. Only when agents give a small importance to the cost (below 0.016), they follow the protocol in most cases (i.e., for more than 80% of the queries to be submitted). We can also see that, even though the proportion of co-utile queries monotonically increases as the cost coefficient decreases, the actual number of co-utile queries starts to decrease for costs coefficients below 0.016. Indeed, when initiator agents give very little importance to the query costs, they tend to submit themselves the queries that do not harm their privacy, rather than using the protocol just to save submission costs.

The privacy of agents is also significantly affected by the lack of co-utility: profile entropies are hardly improved (from the initial average entropy of 2.15) when cost coefficients are high enough (from 1 to 4) to prevent the system from being co-utile in most cases. Interestingly, profile entropies reach a high value (around 3.2) when just 14% of the queries are submitted by following the co-utile protocol (see the privacy evolution of this case in Fig. 6). This shows how even submitting a small amount of *privacy-preserving* queries from other peers has a very positive flattening effect on the responder's profile. In fact, due to the large weight given by the agents to the query costs, the queries they actually submit on behalf of other agents *have to* provide large privacy gains that compensate their submission cost (see Expression (4)).

Table 2 shows the results for Scenario *S3*, in which reputation is also considered by responders as a payoff.

In comparison with the previous scenario, Scenario *S3* results in a significantly greater number of co-utile query submissions for the same coefficients of cost (cost-reputation difference, in this case). The co-utile query proportion among the forwarded queries, in fact, reaches and maintains a value above 0.9 for cost coefficients below 0.063; this is is comparable to the proportion of Scenario *S1*, in which costs were not considered at all. The actual number of queries submitted in a co-utile way is also higher than in Scenario S2, especially for high cost coefficients. Reputation deviations tend to stay steady for cost coefficients below 1.

Table 2 Scenario *S3*: Number and proportion of queries submitted to the WSE following the co-utile protocol, average profile entropy (privacy), and standard deviation of agents' reputations at the end of the simulation, for different cost coefficients

Cost coefficient	Co-utile queries (proportion)	Average profile entropy	Agents' reputation std. deviation
4	7,898 (0.37)	2.80	0.00047
1	8,252 (0.39)	3.16	0.00044
0.25	13,501 (0.64)	3.44	0.00038
0.063	18,986 (0.92)	3.34	0.00038
0.016	18,257 (0.97)	3.10	0.00037
0.004	14,611 (0.95)	2.98	0.00037
0.001	10,899 (0.93)	2.90	0.00037

Table 3 Scenario *S4*: Number and proportion of queries submitted to the WSE following the co-utile protocol, average profile entropy (privacy), and standard deviation of agents' reputations at the end of the simulation, for different cost coefficients

Cost coefficient	Co-utile queries (proportion)	Average profile entropy	Agents' reputation std. deviation
4	14,968 (0.71)	3.11	0.00017
1	16,201 (0.77)	3.30	0.00017
0.25	17,258 (0.82)	3.41	0.00021
0.063	19,671 (0.95)	3.32	0.00026
0.016	18,307 (0.96)	3.13	0.00028
0.004	14,685 (0.95)	3.01	0.00028
0.001	12,090 (0.94)	2.93	0.00029

Finally, Table 3 shows the results for Scenario *S4*, in which the impact of the various available actions on the agent's own reputation is also considered as an atomic utility.

In this case, the interest of the agents to increase their own reputation makes them more motivated to accept query requests from other peers, in spite of query costs or initiators' low reputations. As a result, the system becomes co-utile in most cases and for any cost coefficient values. In fact, the proportions of co-utile queries are quite similar to the ones of scenario *S1*, in which no costs were considered, and the actual numbers of queries submitted in a co-utile way are greater than in any of the previous scenarios.

Moreover, in comparison with Scenario *S3*, in Scenario *S4* we obtain lower standard deviations for the reputation values. Agents are motivated to reach the target reputation but not to exceed it. As a consequence, agents with high enough reputations refuse query requests more frequently than agents with low reputations, which yields a fairer allocation of query costs among the peers.

Finally, it is important to note the beneficial effect of reputation (as a cost-neutralizing feature) on the privacy finally achieved by the agents. In comparison with Scenario *S1* (for which we obtained an average profile entropy of 2.9), with the reputation mechanism we were able to reach substantially higher figures for all cost coefficients, as shown in Fig. 6. Indeed, due to the incurred costs and reputation differences resulting from forwarding or submitting queries, agents become more selective (in terms of the privacy gains) about which queries they forward or accept. As a consequence, the privacy gains of queries are better allocated among the peers.

8 Conclusions

In this chapter, we have shown how to apply the theory of co-utility to enforce rationally sustainable P2P anonymous keyword search, and we have developed several self-enforcing co-utile protocols for that scenario. Moreover, we have discussed the need of a reputation mechanism to introduce artificial incentives that compensate the effect of negative payoffs. Specifically, by means of the reputation mechanism detailed in [8] and in the second chapter of this book (which is itself a co-utile mechanism), we can bring co-utility to many protocols that are not naturally co-utile due to costs incurred by participants [8, 13, 14].

Acknowledgements Funding by the Templeton World Charity Foundation (grant TWCF0095/AB60 "CO-UTILITY") is gratefully acknowledged. Also, partial support to this work has been received from the Government of Catalonia (ICREA Acadèmia Prize to J. Domingo-Ferrer and grant 2014 SGR 537), the Spanish Government (projects TIN2014-57364-C2-1-R "SmartGlacis", TIN2015-70054-REDC and TIN2016-80250-R "Sec-MCloud") and the European Commission (projects H2020-644024 "CLARUS" and H2020-700540 "CANVAS"). The authors are with the UNESCO Chair in Data Privacy, but the views in this work are the authors' own and are not necessarily shared by UNESCO or any of the funding bodies.

References

1. AOL Search Data Mirrors. http://gregsadetsky.com/aol-data/. Accessed 17 Jan 2016
2. Castellà-Roca, J., Viejo, A., Herrera-Joancomartí, J.: Preserving user's privacy in web search engines. Comput. Commun. **32**(13–14), 1541–1551 (2009)
3. Domingo-Ferrer, J., Solanas, A., Castellà-Roca, J.: $h(k)$-Private information retrieval from privacy-uncooperative queryable databases. Online Inf. Rev. **33**(4), 720–744 (2009)
4. Domingo-Ferrer, J., González-Nicolás, Ú.: Rational behavior in peer-to-peer profile obfuscation for anonymous keyword search. Inf. Sci. **185**, 191–204 (2012)
5. Domingo-Ferrer, J., González-Nicolás, Ú.: Rational behavior in peer-to-peer profile obfuscation for anonymous keyword search: the multi-hop scenario. Inf. Sci. **200**, 123–134 (2012)
6. Domingo-Ferrer, J., Martínez, S., Sánchez, D., Soria-Comas, J.: Co-utility: self-enforcing protocols for the mutual benefit of participants. Eng. Appl. Artif. Intell. **59**, 148–158 (2017)
7. Domingo-Ferrer, J., Sánchez, D., Soria-Comas, J.: Co-utility: self-enforcing collaborative protocols with mutual help. Prog. Artif. Intell. **5**(2), 105–110 (2016)
8. Domingo-Ferrer, J., Farràs, O., Martínez, S., Sánchez, D., Soria-Comas, J.: Self-enforcing protocols via co-utile reputation management. Inf. Sci. C **367**, 159–175 (2016)
9. Howe, D.C., Nissenbaum, H.: TrackMeNot: resisting surveillance in web search. In: Lessons from the Identity Trail: Privacy, Anonymity and Identity in a Networked Society, pp. 417–436. Oxford University Press, New York (2009)
10. Murugesan, M., Clifton, C.: Providing privacy through plausible deniable search. In: Proceedings of the SIAM International Conference on Data Mining (SDM 2009), pp. 768–779. SIAM (2009)
11. Reiter, M.K., Rubin, A.D.: Crowds: anonymity for web transactions. ACM Transactions on Information and System Security-TISSEC **1**(1), 66–92 (1998)
12. Sánchez, D., Castellà-Roca, J., Viejo, A.: Knowledge-based scheme to create privacy-preserving but semantically-related queries for web search engines. Inf. Sci. **218**, 17–30 (2013)
13. Sánchez, D., Martínez, S., Domingo-Ferrer, J.: Co-utile P2P ridesharing via decentralization and reputation management. Transp. Res. Part C **73**, 147–166 (2016)
14. Turi, A.N., Domingo-Ferrer, J., Sánchez, D.: Filtering P2P loans on co-utile reputation. In: Proceedings of the 13th International Conference on Applied Computing, AC, pp. 139–146 (2016)
15. Viejo, A., Sánchez, D.: Profiling social networks to provide useful and privacy-preserving web search. J. Association Inf. Sci. Technology **65**(12), 2444–2458 (2014)
16. Viejo, A., Sánchez, D., Castellà-Roca, J.: Preventing automatic user profiling in Web 2.0 applications. Knowl.-Based Syst. **36**, 191–205 (2012)

Co-utile Enforcement of Digital Oblivion

Josep Domingo-Ferrer

Abstract Digital storage in the information society allows perfect and unlimited remembering. Yet, the right of an individual to enforce oblivion for pieces of information about her is part of her fundamental right to privacy. In this chapter, we detail a co-utile solution to digital forgetting based on anonymously fingerprinting expiration dates. Due to co-utility, people who learn information about an individual are rationally interested in helping the individual enforce her oblivion policy. Thanks to this rational involvement, even services for content spreading like Facebook or YouTube would be interested in fingerprinting downloads, thereby effectively enforcing the right of content owners to canceling content.

1 Introduction

The information society rests on the availability of huge, cheap and perfect digital storage. Piles of information about everything and everyone are being gathered continuously and stored digitally. Such a wealth of data can be thought of as an external memory available to all of us: by checking it, we can remember a lot and other people can remember a lot about us. While such perfect and comprehensive remembering may be felt as an advantage when *we* are the ones who remember, it clearly is a shortcoming when it allows *others* to remember things about *us* which we would prefer to be forgotten (embarrassing party pictures, etc.). Perfect remembering may not even be good when we are the ones who remember: e.g. remembering wrongs too well for our entire lifetime may be a hindrance to our happiness. For the above reasons, the right of an individual to enforce oblivion for pieces of information about her is increasingly being regarded as part of her fundamental right to privacy (e.g. see [19] and references therein).

Some potential responses to privacy-encroaching digital remembering are reviewed and criticized in [19]. We briefly summarize this analysis below:

J. Domingo-Ferrer (✉)
UNESCO Chair in Data Privacy, Department of Computer Science and Mathematics, Universitat Rovira i Virgili, Av. Països Catalans 26, 43007 Tarragona, Catalonia, Spain
e-mail: josep.domingo@urv.cat

J. Domingo-Ferrer and D. Sánchez (eds.), *Co-utility*, Studies in Systems, Decision and Control 110, DOI 10.1007/978-3-319-60234-9_5

- *Individual responses*. Digital abstinence is one possible individual response, whereby the individual is assumed to be able to adjust her information sharing behavior. However, getting people to constrain what they wish to share is difficult. Another individual attitude is cognitive adjustment: it consists of living with comprehensive digital memory, while at the same time limiting the influence digital memory has on our decision-making. The idea is to accept that people can change and restrict ourselves to looking at their most recent behavior and opinions. Cognitive adjustment may be too ambitious a goal, because humans cannot help taking into account the information at their disposal.
- *Legal responses*. Information privacy rights enforced by law could limit what information and for how long is stored on other individuals. It is not clear whether the legal system is able to effectively enforce privacy rights and whether the individuals are willing to bring actions against perpetrators.
- *Technology responses*. Enforcing digital rights management (DRM) has been suggested as a way to enable an individual to control her private information [4, 18]. The idea is that individuals would add meta-data to their personal information detailing who can use it, for what purpose and price. Besides technical objections to the very notion of DRM, it seems quite unrealistic to assume that an average individual will spend a few hours detailing usage policies and will keep the above meta-information up-to-date.

The author of [19] proposes expiration dates as a simple solution to enforce digital forgetting. Information stored in digital memory is associated with expiration dates set by the information creator (the user first recording the information). The author goes further to assume that digital storage devices could be made to automatically delete information that has reached or exceeded its expiry date. Alternative approaches based on employing smart cards on the user side to process encrypted content [6] could also be envisioned, whereby the smart card would not decrypt the content after its expiration date.

We adopt here the solution based on expiration dates, but we focus on its secure technical enforcement. First, we need to *embed* in the information the expiration date associated with it, in order to prevent such an expiration date from being altered or suppressed. Second, storage devices which automatically delete information past its expiration date do not currently exist; worse yet, placing trust in the hardware (storage devices, smart cards, etc.) to implement information protection policies has proven to be a flawed approach: e.g., hardware copy prevention mechanisms for CDs and DVDs were easily bypassed.

We specify protocols whereby a content creator can embed an expiration date in the content, spread the content and trace whoever is using and/or transferring the content after the expiration date has passed. We also show how to motivate the players to rationally play their corresponding roles specified in the protocols under the umbrella of co-utility [10, 12] (see also the first chapter of this book). Dishonest receiver tracing is made possible as follows:

- The content carries a different fingerprint for each receiver, so that unlawful content usage and spreading can be traced; we state protocols based on anonymous fingerprinting, whereby the content forwarder does not learn the real identity of honest receivers nor does she see the fingerprinted content these receive (which prevents the forwarder from framing honest receivers by spreading herself content fingerprinted with a receiver's identity);
- The sender does not need to fingerprint and send the content individually to each receiver; doing so would be not only extremely cumbersome, but also quite unrealistic, as we cannot prevent one receiver from forwarding the content to other receivers;
- Since the protocols are co-utile, receivers are rationally interested to collaborate in fingerprinting the content they forward to other interested receivers. With our simple solution, even services for content spreading like Facebook or YouTube would be interested in fingerprinting downloads, thereby effectively enforcing the right of content owners to canceling content. Note that the current Terms of Service of those services [13, 25] only guarantee that they will not further spread the content after the user withdraws it from the service, but they take no responsibility for further spreading conducted by third parties who obtained the content while it was posted.

2 Watermarking, Fingerprinting and Anonymous Fingerprinting

Let $D(0) \in \{0, 1\}^*$ denote some digital content (bit-string) some of whose bits can be changed in such a way that the result remains "close" to $D(0)$ (where "close" means "with a similar utility"), but, without knowing which particular bits were changed, altering a "good portion" of these bits is impossible without rendering the content useless. The changed bits are usually called a mark or watermark; if bits are changed differently for each user receiving the content, the mark can also be called fingerprint. The algorithm used to embed a mark while satisfying the previous two conditions is called a watermarking algorithm; to embed a fingerprint can also be termed "to fingerprint". The second requirement above is actually the marking assumption stated in [2]. Finally, let $\mathscr{D}$ denote the set of all "close copies" of $D(0)$. See [5] for more information on watermarking and fingerprinting.

Whether the original content $D(0)$ is needed for mark/fingerprint recovery depends on the method used for embedding: a watermarking method is said to allow blind detection if only the marked content $\tilde{D}$ and the embedded mark are needed for recovery, in addition to the key used for embedding; methods which need also $D(0)$ are called informed watermarking. In return for their smaller flexibility, informed methods tend to be more robust to content manipulation.

An asymmetric or public-key watermarking method is one in which different keys are used for watermark embedding and recovery, in such a way that the recovery key can be made public without compromising the secrecy of the embedding key. This allows public verifiability of the embedded mark. See [5] for more background on watermarking algorithms.

Anonymous fingerprinting is a special type of fingerprinting in which the fingerprinter does neither see the fingerprinted content nor the receiver's identity except in case of unlawful redistribution. Anonymous fingerprinting procedures were proposed in [1, 3, 7, 8, 22], among others. We take the following model from [3].

Definition 5.1 An anonymous fingerprinting scheme involves a merchant (i.e. a fingerprinter), a buyer (i.e. a receiver) and a registration center. Let c denote the maximal size of a collusion of buyers against which the scheme is secure. An anonymous fingerprinting scheme consists of the following five procedures.

FKG-RC: A probabilistic key setup algorithm for the registration center. Its outputs are the center's secret key x_C and its public key y_C, which is published in an authenticated manner.

FReg: A probabilistic two-party protocol (FReg-RC, FReg-B) between the registration center and the buyer. Their common input is the buyer's identity ID_B and the center's public key y_C. The center's secret input is its secret key x_C. The buyer's output consists of some secret x_B and related information y_B. The center obtains and stores y_B and ID_B.

FPrint: A two-party protocol (FPrint-M, FPrint-B) between the merchant and the buyer. Their common input consists of y_C. The merchant's secret input is $D(0)$ and a transaction number j and her output is a transaction record r_j. The buyer's secret input is x_B and y_B and her output consists of a copy $D(B) \in \mathcal{D}$.

FRec: This may be a protocol or an algorithm whose purpose is to recover the identity of a/the fraudulent buyer responsible for the redistribution of a version $\tilde{D} \in \mathcal{D}$:

- It is a two-party protocol between the merchant and the registration center if the merchant needs the help of the registration center. The merchant's input is a copy $\tilde{D} \in \mathcal{D}$, all transaction records r_i and perhaps the original content $D(0)$. The center's input consists of its secret key x_C and its list of y_B's and ID_B's. The merchant's output is a/the fraudulent buyer's identity together with a proof p that this buyer indeed bought a copy of $D(0)$, or $\perp$ in case of failure (*e.g.*, if more than c buyers colluded to produce $\tilde{D}$).
- It is an algorithm run by the merchant alone if the merchant can determine a/the fraudulent's buyer identity with just $\tilde{D}$, all transaction records r_i and perhaps the original content $D(0)$ (if the underlying watermarking is not blind).

FVer: A verification algorithm, that takes as input the identity ID_B of an accused buyer, the public key y_C of the registration center, and a proof p, and outputs 1 iff the proof is valid.

The solution in [3] guarantees the following properties:

Correctness: All protocols terminate successfully whenever players are honest (no matter how other players behaved in other protocols).

Anonymity and unlinkability: Without obtaining a particular $D(B)$, the merchant –even when colluding with the registration center– cannot identify a buyer (anonymity). Furthermore, the merchant is not able to tell whether two purchases were made by the same buyer (unlinkability).

Protection of innocent buyers: No coalition of buyers, the merchant, and the registration center is able to generate a proof $\tilde{p}$ such that $\mathsf{FVer}(ID_B, y_C, \tilde{p}) = 1$, if buyer ID_B was not present in the coalition.

Revocability and collusion resistance: Any collusion of up to c buyers aiming at producing a version $\hat{D} \in \mathscr{D}$ from which none of them can be re-identified will fail: from $\hat{D}$ the merchant will obtain enough information to identify at least one collusion member.

3 A Protocol Suite for Digital Oblivion

Assume that a player P^0 has a content $D(0)$ for which she wishes to enforce an expiration date $T(0)$. Let $\mathscr{P}^1$ be the set of peers to whom P^0 directly releases a fingerprinted version of $D(0)$. Let $\mathscr{P}^i$ be the set of peers which receive a fingerprinted version of $D(0)$ after i hops, that is, peers in $\mathscr{P}^i$ receive a fingerprinted version of $D(0)$ from some player in $\mathscr{P}^{i-1}$. Note that the sets $\mathscr{P}^i$ are not fixed in advance. The membership of a certain peer P to one of those sets is established when P obtains a fingerprinted version of $D(0)$ from another peer P': if P' belongs to $\mathscr{P}^{i-1}$, then P belongs to $\mathscr{P}^i$. Assuming that the number of hops is limited to N, the following protocol can be used for P^0 to spread $D(0)$ while enforcing expiration date $T(0)$. In the protocol, players are only known by pseudonym, not by their real identities.

Protocol 1

1. *P^0 embeds the expiration date $T(0)$ into $D(0)$ using a blind and robust asymmetric watermarking algorithm such that $T(0)$ can be recovered by anyone seeing the watermarked content but it cannot be altered without substantially damaging the quality of the content. Public recoverability of $T(0)$ is necessary to preclude disputes about the expiration date.*
2. *For any P^{j_1} to whom P^0 wishes to forward the content, P^0 and P^{j_1} run an anonymous fingerprinting scheme such as the one described in Sect. 2 where the underlying watermarking is* blind. *The secret input by P^0 is the watermarked content obtained in Step 1. After running* FReg *and* FPrint, *P^{j_1} obtains a fingerprinted copy $D(0, j_1)$ of the content and P^0 obtains a transaction record $r(0, j_1)$ (partially or totally consisting of information input by P^0 herself like the transaction number).*

3. **for** $i = 1$ **to** $N - 1$ **do**

for all *pairs* $(P^{j_i}, P^{j_{i+1}})$ *such that* $P^{j_i} \in \mathscr{P}^i$ *and* P^{j_i} *wishes to forward the content to* $P^{j_{i+1}}$ **do**

a. Engage in the same anonymous fingerprinting scheme described in Step 2 whereby $P^{j_{i+1}}$ *obtains a fingerprinted version*

$$D(0, j_1, j_2, \ldots, j_{i+1})$$

and P^{j_i} *obtains a transaction record* $r(j_i, j_{i+1})$ *(partially or totally consisting of information input by* P^{j_i} *herself like the transaction number);*

b. P^{j_i} *sends* $r(j_i, j_{i+1})$ *to* P^0.

end for

end for

In Protocol 1 we are implicitly assuming that the asymmetric watermarking scheme used in Step 1 and the watermarking scheme underlying the anonymous fingerprinting scheme allow embedding $T(0)$ and at least the N successive fingerprints corresponding to the N hops in such a way that:

1. It still holds that the resulting $D(0, j_1, \ldots, j_N)$ is close to $D(0)$, that is $D(0, j_1, \ldots, j_N) \in \mathscr{D}$.
2. Embedding a new fingerprint does not destroy the previously embedded fingerprints. In fact, this results from the aforementioned marking assumption and the previous assumption on the "closeness" of $D(0)$ and $D(0, j_1, \ldots, j_N)$.

If the maximum N for which the above holds is chosen, then players will be rationally interested in not performing more than N hops in Protocol 1: indeed, in the same way as more hops would damage the embedded fingerprints, they would damage the perceptual quality of the content, which would no longer belong to $\mathscr{D}$.

The need for blind watermarking and for P^{j_i} to return the transaction record to P^0 will be justified in Sect. 3.1 below.

3.1 Security Analysis

Systematically detecting *all* copies used beyond the expiration date is outside the scope of this work and it is probably an infeasible task in general; e.g. if expired copies are not publicly used, it will be very difficult to spot them. Hence, in this section, we assume that *one* copy of the content $\tilde{D} \in \mathscr{D}$ is detected by P^0 after the expiration date $T(0)$ fingerprinted in it (for example, P^0 finds $\tilde{D}$ posted on the web or some social media). What we target is *oblivion in the public domain*, that is, identifying who makes public use of expired content; private oblivion is certainly impossible to enforce, because each individual's private memories are her own. We focus on how P^0 can identify the redistributor's identity, that is, whose content the detected one is.

P^0 runs the following protocol to establish the identity of the redistributor. Whereas the names P^{j_i} are pseudonyms of the players, the protocol allows finding real identities of players. For $i = 1$ to N, let R^i be the set of i-hop transaction records received by P^0 in Protocol 1 for $i = 1$ up to N, that is,

$$R^i = \{r(j_{i-1}, j_i) : \forall P^{j_{i-1}} \in \mathcal{P}^{i-1}, P^{j_i} \in \mathcal{P}^i\},$$

where $P^{j_0} := P^0$ and $r(j_0, j_1) := r(0, j_1)$. Further, let $R^i(P^{j_{i-1}})$ be the subset of i-hop transaction records originated by player $P^{j_{i-1}}$, that is,

$$R^i(P^{j_{i-1}}) = \{r(j_{i-1}, j_i) : \forall P^{j_i} \in \mathcal{P}^i\}.$$

Protocol 2

1. *Set* $i := 1$ *and* $finished :=$ **false**;
2. **while** $finished =$ **false and** $i \leq N$ **do**

   if $R^i(P^{j_{i-1}}) \neq \emptyset$ **then**

   a. *Sample without replacement one transaction record* r' *from* $R^i(P^{j_{i-1}})$;
   b. *Run with the registration center the* FRec *protocol with inputs the redistributed content* $\tilde{D}$ *and* r';
   c. **if** *the real identity of a player* P^{j_i} *can be successfully recovered* **then**
      i. *Request the set* $R^{i+1}(P^{j_i})$ *from* P^{j_i};
      ii. $i := i + 1$;

   else *finished*:= **true**;

   end while

3. *Output the real identity of* $P^{j_{i-1}}$ *as the redistributor's identity.*

Some remarks on the above identity recovery process follow:

- The anonymous fingerprinting scheme used must be based on an underlying blind watermarking method. Indeed, by construction of Protocol 1, unless the redistributor belongs to $\mathcal{P}^1$, P^0 does not know the original unmarked content corresponding to $\tilde{D}$. Imagine that the redistributor is some $P^{j_i} \in \mathcal{P}^i(P^{j_{i-1}})$ where $\mathcal{P}^i(P^{j_{i-1}})$ is the set of i-hop players who received the content from $P^{j_{i-1}}$, with $1 < i \leq N$. In that case, $\tilde{D} = D(0, j_1, \ldots, j_i)$ and the original unmarked content is $D(0, j_1, \ldots, j_{i-1})$, only known to $P^{j_{i-1}}$, not to P^0.
- Unless a player redistributes her received content, she is the only one to know that content, because anonymous fingerprinting is also asymmetric: the fingerprinter does not see the fingerprinted content. Therefore, no one can frame an honest receiver P^{j_i} by accusing her of redistributing $D(0, j_1, \ldots, j_i)$, because the latter content is known to no one but P^{j_i}.

- P^0 only wants to obtain the identity of the *last* player who fingerprinted the redistributed content. There is a trade-off between anonymity preservation and search efficiency:
  - Since spreading is potentially exponential, Protocol 2 tries to reduce the search space and the computational burden for P^0, even if this implies anonymity loss for all players along the path between P^0 and the redistributor.
  - Since P^0 receives all transaction records in Protocol 1, if preserving the anonymity of all honest players is more important than reducing the search space, Protocol 2 could be modified for P^0 to start searching backwards: first search $\mathcal{P}^N$ by trying FRec with the redistributed content and all N-hop transaction records; if the redistributor was not found, then search $\mathcal{P}^{N-1}$, etc.; the search would stop as soon as a real identity (the redistributor's) could be successfully recovered. The first and only recovered identity would be the redistributor's; the registration center would refuse its necessary collaboration in FRec if P^0 tried to identify other players after the redistributor had been found. This approach prioritizes anonymity preservation but it can implode P^0, who must try a virtually exponential number of transaction records.
- Protocol 2 works even if some peers fail to send the transaction record to P^0 in Protocol 1. Assume that P^{j_i} and $P^{j_{i+1}}$ engage on a transfer of content and that this transfer takes place without any fingerprinting or without P^{j_i} sending the resulting transaction record to P^0. Since P^0 does not have the transaction record for the transfer to $P^{j_{i+1}}$, P^0 will not be able to identify $P^{j_{i+1}}$ in case $P^{j_{i+1}}$ performs unauthorized redistribution. By following the chain of transaction records, Protocol 2 will accuse P^{j_i} of redistribution. Note that P^{j_i} is indeed guilty for not having correctly followed Protocol 1 and thus can be held liable for the redistribution performed by $P^{j_{i+1}}$ or by anyone in $\mathcal{P}^{i+2}(P^{j_{i+1}})$. A particular case of the above is when P^{j_i} publishes the content on the web or otherwise releases it without keeping track of who accesses it. In that case, P^{j_i} will be accused of redistribution.

3.2 Practical Instantiation

Blind and robust asymmetric watermarking schemes for Step 1 of Protocol 1 can be found, for example, in [14–16, 24]. The audio watermarking scheme described in [20] can be used as a scheme underlying the anonymous fingerprinting in Protocol 1, because it satisfies the requirements listed above. The scheme is blind, so that it is possible to extract the embedded mark from a marked audio object without knowing the original unmarked audio object.

Also, the scheme tolerates embedding several successive fingerprintings without significant damage to the content utility or the previous fingerprints. We have used the Objective Difference Grade (ODG) based on the ITU-R Recommendation standard BS 1387 [17, 23] to evaluate the damage inflicted by successive fingerprintings. This standard makes it possible to evaluate the transparency of the fingerprinting

Table 1 Imperceptibility results with five successive fingerprintings

Content	# fingerprints	ODG
$D(0, j_1)$	1	0.000
$D(0, j_1, j_2)$	2	−0.004
$D(0, j_1, j_2, j_3)$	3	−0.034
$D(0, j_1, j_2, j_3, j_4)$	4	−0.115
$D(0, j_1, j_2, j_3, j_4, j_5)$	5	−0.193

scheme by comparing the perceptual quality of the marked files with respect to the original content $D(0)$. The ODG values are in the range $[-4, 0]$, where 0 means imperceptible, -1 means perceptible but not annoying, -2 means slightly annoying, -3 means annoying and -4 means very annoying. In order to evaluate the ODG, we have used the Opera software by Opticom [21]. The imperceptibility results are shown in Table 1. As it can be noticed, the transparency slowly decreases for each successive receiver in Protocol 1. However, even with 5 embedded fingerprints, the ODG result is much closer to 0 (imperceptible) than -1 (perceptible but not annoying); hence, even in this worst case, the perceptual quality achieved by Protocol 1 with the scheme [20] can be regarded as very satisfactory.

Furthermore, the scheme can be adapted for anonymous fingerprinting. The adaptation is described in detail in [11] and summarized next. The scheme uses a double embedding strategy:

- A time-domain synchronization watermark ("SYN") is embedded for fast search of the information watermark position;
- A frequency-domain information watermark is embedded next to the SYN marks.

This double embedding strategy makes it possible to embed the transaction records $r(j_i, j_{i+1})$ and the receiver related information y_B in different domains. The transaction records can be embedded as synchronization marks in the time domain with different bit strings, and the related information y_B can be embedded more robustly in the frequency domain. This scheme has the additional advantage of a very fast search of transaction records; extracting an embedded transaction record from a portion of audio takes less time than playing that portion.

In order to preserve anonymity and make the registration center necessary for redistributor identification (as mandated by Protocol 2), we let y_B be the receiver identity encrypted under the registration center's public key. To obtain unlinkability, a random nonce is appended to the receiver's identity before encrypting it under the registration center's public key. Embedding next to y_B a hash of x_B (or x_B encrypted with the public key of the receiver) has the additional advantage of thwarting a collusion of the sender P^0 and the registration center, who would not be able to produce a correctly fingerprinted copy of the content corresponding to any receiver.

If the sender P^0 finds a version of the audio file illegally redistributed on the Internet, she can search for the transaction records in the time domain (fast search) and then extract the information y_B related to the malicious receiver. This information (y_B) will then be sent to the registration center in order to identify the illegal redistributor.

We published the scheme described so far in [9]. Next, we justify its co-utility.

4 Rational Involvement of Players by Means of Co-utility

In this section, we show how to motivate the players to rationally play their corresponding roles specified in the protocols under the umbrella of co-utility. Showing that players have no interest in deviating is especially necessary in peer-to-peer (P2P) protocols whose correct operation depends on the commitment of the players. For more details on the notion and formal definition of co-utility, see the first chapter of this book (or [10, 12]).

P^0 has an obvious interest in correctly following Protocol 1. If she deviates by not correctly participating in the anonymous fingerprinting process, the entire expiration date enforcement does not even start. Let s^0 be the strategy whereby P^0 follows Protocol 1. Let us now analyze the strategies of the other players.

Each player $P^{j_i} \neq P^0$ has at least the following possible strategies with respect to P^0 and $P^{j_{i+1}} \in \mathscr{P}^{i+1}(P^{j_i})$:

- $s_0^{j_i}$: Correctly follow Protocol 1 by engaging in anonymous fingerprinting with $P^{j_{i+1}}$ and returning transaction record $r(j_i, j_{i+1})$ to P^0;
- $s_1^{j_i}$: Deviate from Protocol 1 by engaging in anonymous fingerprinting with $P^{j_{i+1}}$ but *not* returning $r(j_i, j_{i+1})$ to P^0;
- $s_2^{j_i}$: Deviate from Protocol 1 by anonymously forwarding to $P^{j_{i+1}}$ the content without fingerprinting and returning a fake transaction record $r(j_i, j_{i+1})$ to P^0;
- $s_3^{j_i}$: Deviate from Protocol 1 by anonymously forwarding the content without fingerprinting and not returning any transaction record;
- $s_4^{j_i}$: Deviate from Protocol 1 by not forwarding the content and not returning any transaction record.

We next discuss how to achieve co-utility, that is, how Protocols 1 and 2 need to be modified to ensure that, for any player P^{j_i}, her rational choice is strategy $s_0^{j_i}$.

4.1 *Utility Without Reward or Punishment*

Consider the following payoffs:

- d_{j_i}: Payoff that P^{j_i} derives from obtaining $D(0, j_1, \ldots j_i)$, *without losing her anonymity*. That is, d_{j_i} combines the functionality payoff of P^{j_i} obtaining the content and the privacy payoff of P^{j_i} preserving her anonymity thanks to anonymous fingerprinting with $P^{j_{i-1}}$. If P^{j_i} pays a fee or reward for obtaining the content, this fee or reward must be deducted from d_{j_i}.

$-v_{j_i}$: Negative payoff (that is, cost) that P^{j_i} incurs from engaging in anonymous fingerprinting with $P^{j_{i+1}}$; this cost may be quantified in terms of computation and communication.

$-w_{j_i}$: Negative payoff (that is, communication cost) that P^{j_i} incurs for returning the transaction record $r(j_i, j_{i+1})$ to P^0.

$-f_{j_i}$: Negative payoff (communication cost) that P^{j_i} incurs from anonymously forwarding the unfingerprinted content to $P^{j_{i+1}}$. Obviously $f_{j_i} \leq v_{j_i}$ because forwarding unfingerprinted content is simpler than anonymously fingerprinting it.

If there are no other payoffs (like reward earned for following the protocol or punishment incurred for not following it), the general utility functions of the above strategies are the following:

$$\mathbf{u}_{j_i}(s_0^{j_i}) = d_{j_i} - v_{j_i} - w_{j_i};$$

$$\mathbf{u}_{j_i}(s_1^{j_i}) = d_{j_i} - v_{j_i};$$

$$\mathbf{u}_{j_i}(s_2^{j_i}) = d_{j_i} - f_{j_i} - w_{j_i};$$

$$\mathbf{u}_{j_i}(s_3^{j_i}) = d_{j_i} - f_{j_i};$$

$$\mathbf{u}_{j_i}(s_4^{j_i}) = d_{j_i}.$$

Clearly, strategy $s_4^{j_i}$ has the maximum utility. Considering a sequence of players $P^0, P^{j_1}, \ldots, P^{j_N}$ with $P^{j_i} \in \mathscr{P}^i(P^{j_{i-1}})$ for $i = 1$ to N, the dominant strategy solution of the game is

$$(s^0, s_4^{j_1}, -, \ldots, -).$$

In plain words, the rational equilibrium is for P^0 to conduct anonymous fingerprinting of the content with P^{j_1} and for P^{j_1} to acquire the anonymously fingerprinted content and do nothing else. The strategies of the rest of players are irrelevant, because they receive no content.

4.2 *Utility with Reward and No Punishment*

In an attempt to induce rational players to correctly follow Protocol 1, we can think of introducing a reward for a player who forwards the content to other players. $P^{j_{i+1}}$ pays a reward g_{j_{i+1},j_i} to P^{j_i} upon receiving the content. $P^{j_{i+1}}$ discounts g_{j_{i+1},j_i} from her payoff $d_{j_{i+1}}$.

In this case, the utility functions of the four strategies of P^i are:

$$\mathbf{u}_{j_i}(s_0^{j_i}) = d_{j_i} + g_{j_{i+1},j_i} - v_{j_i} - w_{j_i};$$

$$\mathbf{u}_{j_i}(s_1^{j_i}) = d_{j_i} + g_{j_{i+1},j_i} - v_{j_i};$$

$$\mathbf{u}_{j_i}(s_2^{j_i}) = d_{j_i} + g_{j_{i+1},j_i} - f_{j_i} - w_{j_i};$$

$$\mathbf{u}_{j_i}(s_3^{j_i}) = d_{j_i} + g_{j_{i+1},j_i} - f_{j_i};$$

$$\mathbf{u}_{j_i}(s_4^{j_i}) = d_{j_i}.$$

If the reward is sufficient to cover the costs of P^{j_i} anonymously forwarding the content without fingerprinting, that is, if $g_{j_{i+1},j_i} \geq f_{j_i}$, then $s_3^{j_i}$ has the maximum utility. In these conditions, the dominant strategy solution of the game is

$$(s^0, s_3^{j_1}, \ldots, s_3^{j_{N-1}}, s_4^{j_N}).$$

In plain words, the rational equilibrium is for P^0 to correctly follow Protocol 1 with P^{j_1} and for P^{j_i} $(i = 1, \ldots, N-1)$ to anonymously forward the content to $P^{j_{i+1}}$ without fingerprinting and without returning any transaction records to P^0. The strategy of P^N can only be $s_4^{j_N}$, because P^{j_N} is not supposed to forward the content any further.

4.3 Utility with Reward and Punishment

A punishment mechanism is needed to motivate players P^{j_1} through $P^{j_{N-1}}$ to return valid transaction records to P^0 (and therefore correctly engage in anonymous fingerprinting with the next player).

Let $-p_{j_i}$ be the expected negative payoff (punishment) that P^{j_i} incurs when accused of redistribution as a result of not having returned a valid transaction record $r(j_i, j_{i+1})$. This is actually an *expected* negative payoff, computed as the probability of being accused times the cost of being accused: such cost may be a fine, the utility loss of being disgraced vs P^0, etc. We can recompute the utilities of the four strategies available to P^{j_i}:

$$\mathbf{u}_{j_i}(s_0^{j_i}) = d_{j_i} + g_{j_{i+1},j_i} - v_{j_i} - w_{j_i};$$

$$\mathbf{u}_{j_i}(s_1^{j_i}) = d_{j_i} + g_{j_{i+1},j_i} - v_{j_i} - p_{j_i};$$

$$\mathbf{u}_{j_i}(s_2^{j_i}) = d_{j_i} + g_{j_i,j_{i+1}} - f_{j_i} - w_{j_i} - p_{j_i}; \quad (1)$$

$$\mathbf{u}_i(s_3^i) = d_{j_i} + g_{j_{i+1},j_i} - f_{j_i} - p_{j_i};$$

$$\mathbf{u}_i(s_4^i) = d_{j_i}.$$

Assume like above that $g_{j_{i+1},j_i} \geq f_{j_i}$. Assume also that

$$v_{j_i} + w_{j_i} < p_{j_i}, \tag{2}$$

that is, anonymously fingerprinting the content and returning the correct transaction record is less costly than the punishment. Then,

$$\mathbf{u}_{j_i}(s_1^{j_i}) \leq \mathbf{u}_{j_i}(s_0^{j_i});$$

$$\mathbf{u}_{j_i}(s_2^{j_i}) \leq \mathbf{u}_{j_i}(s_0^{j_i});$$

$$\mathbf{u}_{j_i}(s_3^{j_i}) \leq \mathbf{u}_{j_i}(s_0^{j_i}).$$

With the above constraints, P^{j_i} will rationally choose $\mathbf{u}_{j_i}(s_0^{j_i})$ if the reward is sufficient to cover the costs of anonymous fingerprinting and returning the transaction record, i.e. if

$$g_{j_i,j_{i+1}} \geq v_i + w_i. \tag{3}$$

Otherwise, P^{j_i} will choose $\mathbf{u}_{j_i}(s_4^{j_i})$ (no forwarding). At any rate, the strategies consisting of forwarding unfingerprinted content will not be chosen by P^{j_i}. Therefore, even if $P^{j_{i+1}}$ has no rational interest in getting a fingerprinted content, she will have to increase the reward to meet Inequality (3) if she wants to be forwarded the content at all. Under conditions (2) and (3), the dominant strategy solution of the game is

$$(s^0, s_0^{j_1}, \ldots, s_0^{j_{N-1}}, s_3^{j_N}).$$

In plain words, with the proposed modifications, we have succeeded in inducing a rational behavior for P^0 and players P^{j_i} $(i = 1, \ldots, N-1)$ which consists of correctly following Protocol 1. The strategy of P^{j_N} can only be $s_4^{j_N}$, because P^{j_N} is not supposed to forward the content any further.

Lemma 5.1 *With the utility functions defined in Eq. (1) and the constraints (2) and (3), in Protocol 1 there is co-utility between P^{j_i} and $P^{j_{i+1}}$ for any $i \in \{0, \ldots N-1\}$. Additionally, the equilibrium strategy between P^{j_i} and $P^{j_{i+1}}$ results in increased redistribution tracing capabilities for P^0.*

Proof With the utilities in this section, the dominant strategy solution has been shown to be the one in which every player P^{j_i} plays $s_0^{j_i}$. Note that $s_0^{j_i}$ yields the maximal possible payoff $d_{j_{i+1}}$ for $P^{j_{i+1}}$: indeed, $P^{j_{i+1}}$ obtains the content while preserving her anonymity (thanks to anonymous fingerprinting). Now, whatever the strategy chosen by $P^{j_{i+1}}$, the general utility function $\mathbf{u}_{j_{i+1}}$ monotonically increases with $d_{j_{i+1}}$.

Hence, the best strategy for P^{j_i} results in enhanced general utility for $P^{j_{i+1}}$, whatever $P^{j_{i+1}}$'s strategy.

Additionally, since the best strategy for P^{j_i} is to transfer fingerprinted content, P^0 increases her tracing capability, because she can trace this transfer in case of redistribution (Protocol 2). The lemma follows.

5 Conclusions

We have described protocols whereby expiration dates for digital content can be enforced thanks to rational involvement by all players having received the content. This rational, co-utile approach allows digital oblivion to be effectively enforced.

Acknowledgements Funding by the Templeton World Charity Foundation (grant TWCF0095/AB60 "CO-UTILITY") is gratefully acknowledged. Also, partial support to this work has been received from the Government of Catalonia (ICREA Acadèmia Prize to J. Domingo-Ferrer and grant 2014 SGR 537), the Spanish Government (projects TIN2014-57364-C2-1-R "SmartGlacis", TIN2015-70054-REDC and TIN2016-80250-R "Sec-MCloud") and the European Commission (projects H2020-644024 "CLARUS" and H2020-700540 "CANVAS"). The authors are with the UNESCO Chair in Data Privacy, but the views in this work are the authors' own and are not necessarily shared by UNESCO or any of the funding bodies.

References

1. Bo, Y., Piyuan, L., Wenzheng, Z.: An efficient anonymous fingerprinting protocol. In: Computational Intelligence and Security, LNCS 4456, pp. 824–832. Springer, Berlin (2007)
2. Boneh, D., Shaw, J.: Collusion-secure fingerprinting for digital data. In: Advances in Cryptology, LNCS 963, pp. 452–465. Springer, Berlin (1995)
3. Camenisch, J.: Efficient anonymous fingerprinting with group signatures. In: Asiacrypt, LNCS 1976, pp. 415–428. Springer, Berlin (2000)
4. Cohen, J.E.: DRM and privacy. Berkeley Technol. Law J. **18**, 575–617 (2003)
5. Cox, I.J., Miller, M.L., Bloom, J.A., Fridrich, J., Kalker, T.: Digital Watermarking and Steganography. Morgan Kaufmann, Burlington MA (2008)
6. Domingo-Ferrer, J.: Multi-application smart cards and encrypted data processing. Future Gen. Comput. Syst. **13**(1), 65–74 (1997)
7. Domingo-Ferrer, J.: Anonymous fingerprinting of electronic information with automatic identification of redistributors. Electron. Lett. **34**(13), 1303–1304 (1998)
8. Domingo-Ferrer, J.: Anonymous fingerprinting based on committed oblivious transfer. In: Public Key Cryptography-PKC 1999, LNCS 1560, pp. 43–52. Springer, Berlin (1999)
9. Domingo-Ferrer, J.: Rational enforcement of digital oblivion. In: Proceedings of the 4th International Workshop on Privacy and Anonymity in the Information Society, ACM, Art. no. 2 (2011)
10. Domingo-Ferrer, J., Martínez, S., Sánchez, D., Soria-Comas, J.: Co-utility: self-enforcing protocols for the mutual benefit of participants. Eng. Appl. Artif. Intell. **59**, 148–158 (2017)
11. Domingo-Ferrer, J., Megías, D.: Distributed multicast of fingerprinted content based on a rational peer-to-peer community. Comput. Commun. **36**, 542–550 (2013)
12. Domingo-Ferrer, J., Sánchez, D., Soria-Comas, J.: Co-utility: self-enforcing collaborative protocols with mutual help. Prog. Artif. Intell. **5**(2), 105–110 (2016)
13. Facebook Terms of Service, revision dated 30 Jan 2015. Checked 2017/05/18 11:59:56. http://www.facebook.com/terms.php

14. Furon, T., Duhamel, P.: An asymmetric public detection watermarking technique. In: Information Hiding IH'99, LNCS 1768, pp. 88–100. Springer, Berlin (1999)
15. Furon, T., Duhamel, P.: An asymmetric watermarking method. IEEE Trans. Signal Process. **51**(4), 981–994 (2003)
16. Gui, G.-F., Jiang, L.-G., He, C.: A robust asymmetric watermarking scheme using multiple public watermarks. IEICE Trans. Fundam. **E88-A**(7):2026–2029 (2005)
17. ITU-R.: Recommendation BS.1387. Method for Objective Measurements of Perceived Audio Quality (1998)
18. Mayer-Schönberger, V.: Beyond copyright: managing information rights with DRM. Denver Univ. Law Rev. **84**, 181–198 (2006)
19. Mayer-Schönberger, V.: The Virtue of Forgetting in the Digital Age. Princeton University Press, Princeton (2009)
20. Megías, D., Serra-Ruiz, J., Fallahpour, M.: Efficient self-synchronised blind audio watermarking system based on time domain and FFT amplitude modification. Signal Process. **90**(12), 3078–3092 (2010)
21. Opticom. Opera software (2015). http://www.opticom.de/products/opera.html. Accessed 18 Nov 2015
22. Pfitzmann, B., Waidner, M.: Anonymous fingerprinting. In: Advances in Cryptology-EUROCRYPT'96, LNCS 1233, pp. 88–102. Springer, Berlin (1997)
23. Thiede, T., Treurniet, W.C., Bitto, R., Schmidmer, C., Sporer, T., Beerends, J.G., Colomes, C., Keyhl, M., Stoll, G., Brandeburg, K., Feiten, B.: PEAQ - the ITU standard for objective measurement of perceived audio quality. J. Audio Eng. Soc. **48**(1–2), 3–29 (2000)
24. Yan-Jun, H., Xiao-Ping, M., Li, G.: A robust public-key image watermarking scheme based on weakness signal detection using chaos system. In: International Conference on Cyberworlds, pp. 477–480 (2009)
25. YouTube Terms of Service, revision dated 9 June 2010. http://www.youtube.com/static?gl=US&template=terms. Accessed 20 Feb 2017

Co-utile Privacy-Aware P2P Content Distribution

David Megías and Josep Domingo-Ferrer

Abstract Multicast distribution of content is not suited to content-based electronic commerce because all buyers obtain exactly the same copy of the content, in such a way that unlawful redistributors cannot be traced. Unicast distribution has the shortcoming of requiring one connection with each buyer, but it allows the merchant to embed a different serial number in the copy obtained by each buyer, which enables redistributor tracing. Peer-to-peer (P2P) distribution is a third option which may combine some of the advantages of multicast and unicast: on the one hand, the merchant only needs unicast connections with a few seed buyers, who take over the task of further spreading the content; on the other hand, if a proper fingerprinting mechanism is used, unlawful redistributors of the P2P distributed content can still be traced. In this chapter, we describe a co-utile fingerprinting mechanism for P2P content distribution which allows redistributor tracing, while preserving the privacy of most honest buyers and offering collusion resistance and buyer frameproofness.

1 Introduction

If a content is to be distributed to a group of N receivers, one option is for the content sender to engage in N unicast transmissions, one for each intended receiver, and another option is a single multicast transmission to the entire group. Certainly, the multicast option has the advantage of being faster and more bandwidth-efficient from the senders point of view. However, the unicast approach has the strong point of allowing the sender to fingerprint the content sent to each receiver by embedding a different serial number in each sent copy, with the aim of detecting and tracing

D. Megías (✉)
Internet Interdisciplinary Institute (IN3), Universitat Oberta de Catalunya,
Av. Carl Friedrich Gauss 5, 08860 Castelldefels, Catalonia, Spain
e-mail: david.megias@uoc.edu

J. Domingo-Ferrer
UNESCO Chair in Data Privacy, Department of Computer Science and Mathematics,
Universitat Rovira i Virgili, Av. Països Catalans 26, 43007 Tarragona, Catalonia, Spain
e-mail: josep.domingo@urv.cat

J. Domingo-Ferrer and D. Sánchez (eds.), *Co-utility*, Studies in Systems,
Decision and Control 110, DOI 10.1007/978-3-319-60234-9_6

unlawful redistribution of the content. Note that the multicast approach does not allow fingerprinting, as all receivers obtain exactly the same content. Hence, the unicast approach, in spite of its inefficiency, seems more suitable when the sender is a merchant selling content and the receivers are buyers.

Peer-to-peer (P2P) distribution of content appears as a third option blending some of the advantages of the unicast and multicast solutions. P2P distribution of all types of files and contents has become extremely popular with the increased bandwidth of home Internet access in the last few years. In addition, P2P file sharing applications are not restricted to this use, and some companies are also exploiting the P2P distribution paradigm as a way of saving server bandwidth and speeding up the downloads of their products (such as multimedia contents and software updates). Indeed, when using a P2P network for content distribution, the merchant only needs to establish direct connections with one or a few seed buyers, say $M \ll N$ buyers, and send them copies. The content is further spread over the P2P network by those seed buyers. The challenge is how to ensure that the P2P spread content is still traceable in case of redistribution.

The type of fingerprinting relevant to this study is anonymous fingerprinting. In anonymous fingerprint schemes, the merchant does not have access to the identities or the fingerprints of buyers, which protects their security and privacy. Initial anonymous fingerprinting proposals depended on unspecified multiparty secure computation protocols [7, 25]. In [8], an anonymous fingerprinting protocol completely specified from the computational point of view and based on committed oblivious transfers was described. In [9], anonymous fingerprinting protocols were simplified under the assumption that a tamper-proof smart card was available on the buyer's side.

Many anonymous fingerprinting schemes exploit some homomorphic property of public-key cryptography [17, 18, 23, 26, 28]. These schemes allow embedding the fingerprint in the encrypted domain (using the public key of the buyer) in such a way that only the buyer obtains the decrypted fingerprinted content. However, developing a practical system using these ideas appears difficult, because public-key encryption expands data and substantially increases the communication bandwidth required for transfers [16]. In [3], a different approach using group signatures was suggested, but this solution requires bit commitment and a zero-knowledge proof, implying a large overhead and high communicational costs. In the proposal of [1], the system's efficiency is enhanced due to the suppression of zero-knowledge proofs and public-key cryptography is not required in the embedding scheme. However, a secure two-party computation protocol is used between the merchant and each buyer to transfer the fingerprinted content. In [16], any secure watermarking scheme (for which no proof of existence is available) may be used to develop an anonymous fingerprinting protocol if the watermark embedder provides a certain level of security. Although the proposed approach avoids the costs of homomorphic cryptography, a practical application of that idea is not presented. Another proposal to reduce the burden of anonymous fingerprinting on the buyer's side is presented in [4], where powerful servers would perform the most costly parts of the protocols. In any case, all the proposed anonymous fingerprinting systems incur high computational and communicational burdens at the buyer's and/or at the merchant's side, due to the use of some highly demanding technology (public-key encryption of the contents, secure multiparty protocols or zero-knowledge proofs, among others). Some of them also require specific embedding schemes which are not among the most robust or secure ones, or a secure watermarking system that is not proven to exist. In this study, we propose a novel solution to overcome these drawbacks, since the use of public-key cryptography is restricted to the transmission of short bit strings (hashes) and is not applied to the multimedia content itself. In addition, the proposed scheme decentralizes the transmission of the content using a network of peer buyers, thereby reducing the bandwidth needed by the merchant.

We propose a co-utile [10, 12] P2P distribution scheme of fingerprinted content whereby the merchant originates only a set of M seed copies of the content and sends them to M seed buyers. All subsequent copies are generated from the seed copies. Each non-seed buyer obtains her copy of the content by running a P2P purchase software tool. The copy obtained by each buyer is a combination of the copies provided by her sources (parents). The fingerprint of each buyer is thus a binary sequence *automatically* formed as the combination of the sequences of her sources. This peer-to-peer distribution scheme makes it possible for the merchant to save bandwidth and CPU time, while still being able to trace unlawfully redistributed content. Moreover, the co-utility property ensures that, whenever a child buyer can obtain her content from more than one parent, she will do so; it also ensures that parents will be interested in not passing their entire content to a single child buyer (for details on co-utility, see the first chapter of this book and/or the references cited at the beginning of this paragraph).

2 Overview of the Proposed Scheme

In the proposed P2P scheme for distributing fingerprinted content (see end of previous section), the fingerprints of the buyers do not need to be registered in any way and, thus, all buyers can preserve their privacy as long as no illegal content redistribution occurs. However, when an illegally redistributed file is found, it is possible to link its binary sequence to a particular individual (buyer). As in most fingerprinting applications, in the proposed system "illegal redistribution" means that the buyer redistributes the whole content (file) to a third party who does not purchase it legally, or makes the content available for download in a non-authorized platform (web page, file sharing application, or other) without the copyright owner's explicit permission. The tracing of an illegal redistributor does not need to be particularly fast and legal actions can be taken when the identification is completed.

To satisfy the above conditional privacy, a P2P proxy (or set of proxies) is used to create anonymous connections between buyers such that source and destination buyers do not lose their anonymity. The P2P proxy also sends a transaction record to a transaction monitor whenever a buyer obtains fragments of the contents from another buyer. The fields of this transaction record are the following:

- An identifier of the purchased contents (a perceptual hash).
- The pseudonyms of the two buyers participating in the transaction, that is, the parent and the child.
- The encrypted fingerprint of the contents obtained by the child from the parent.
- The encrypted hash of the fingerprint of the contents obtained by the child from the parent.
- The time and date of the transaction.

A child is supposed to obtain pieces of the content from several parents, so there will be one transaction record for each parent the child gets pieces from. The purpose of storing the above transaction records at the transaction monitor is to enable tracing of illegal redistributors.

Buyers stay anonymous to each other, but only pseudonymous versus the transaction monitor; however, the transaction record does not specify which fragments come from which buyer, so that the privacy of the buyers' fingerprints is preserved. The encrypted hash is used by the authority in case a buyer intends to cheat the tracing system by showing a different (modified or borrowed) copy of the content. Since the transaction monitor only records a hash of the true fingerprint and buyer pseudonyms that are not linked to specific fragments of the content, no coalition of the transaction monitor, the seller or other buyers can be used to frame an innocent buyer (by unjustly accusing her).

In order to carry out an a posteriori identification of redistributors, a basic traitor protocol is run taking the fingerprint of the illegally redistributed content. In addition, the registered hashes of the fingerprints are enough to discourage buyers from cheating the tracing system by using borrowed or altered copies of the contents. On the other hand, the privacy of the honest users is preserved and their fingerprints remain private.

In addition to attractive privacy properties, it will be shown that, in practice, the proposed scheme offers good security properties, namely collusion resistance versus dishonest colluding buyers (if a particular anti-collusion strategy is used) and buyer frameproofness versus a malicious merchant.

3 P2P Distribution of Recombined Fingerprinted Contents

The basis of P2P content distribution is that the shared contents are distributed by some users to others. As soon as some fragments of the content are received, destination users become sources for others. A file is thus obtained by joining the fragments of several sources together. Typically, a hash value of the shared content is used by P2P clients to identify files. Two files having the same hash value are considered equal. The upload/download process of a file from different sources is depicted in Fig. 1. In this figure, the destination obtains fragments from three different sources that are joined together to form the content.

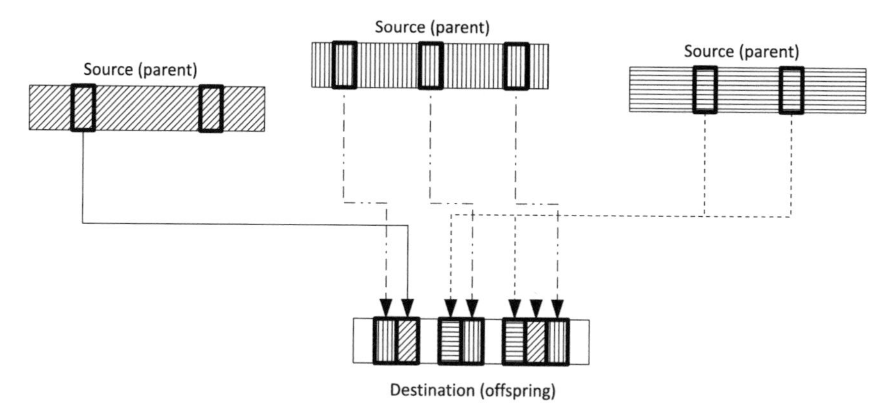

Fig. 1 Upload/download of the content (mating process)

3.1 Mating Approach for Fingerprinting

In this section, we introduce a novel concept of automatic binary fingerprints partly inspired on biological mating and inheritance. The relationship between biology and the scheme in this work is rather weak and, thus, we refrain from calling our scheme "genetic". However, some biological analogies are highlighted in this section to introduce the basis of the suggested scheme.

In this work, fingerprints are constructed as binary sequences and each bit might be considered as the counterpart of the nucleotides of a DNA sequence. This is similar to the approach taken in Genetic Algorithms [15] for solving optimization problems. Just like DNA sequences are formed by different genes which encode a given protein, the binary fingerprints used in this work are formed by (fixed-size) segments that may be considered as analogs of genes. When a buyer obtains a copy of a P2P-distributed content using some specific software, the binary fingerprint of her copy is a combination of the segments of the sources of the content (referred to as "parents" from the biological analogy). In this case, the number of parents of some content does not have to be exactly two as in the natural world. Hence, the mating process in the suggested fingerprinting scenario must be understood in a generalized sense, not limited to two parents. Fingerprints can be considered as being "automatically generated" from the fingerprints of the parents. Despite this "automatic generation" of fingerprints, the constructed sequences are still valid for identification purposes. In order to identify the culprit of an illegal redistribution, a identification process must be carried out. The identification process is performed with the help of the fingerprint of the illegally re-distributed copy.

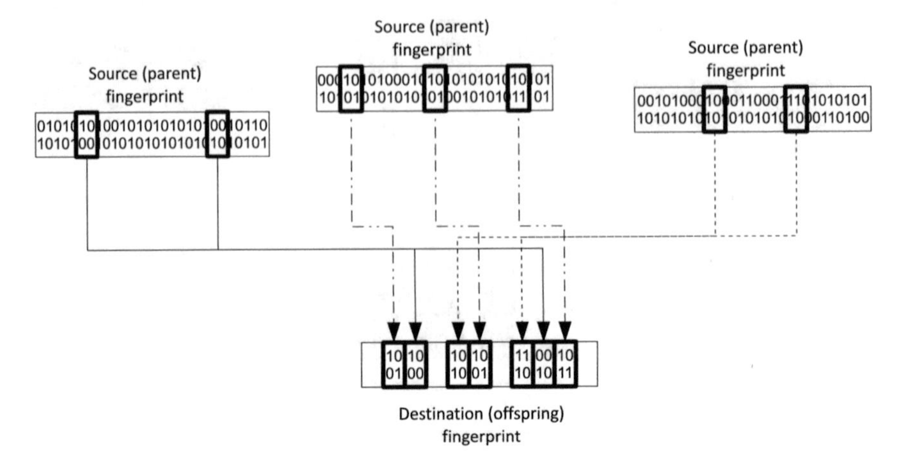

Fig. 2 Automatic recombined fingerprint construction

3.2 Requirements on Fingerprint Embedding

If the binary fingerprints described in Sect. 3.1 are to be found in a P2P-distributed content, an embedding method is to be used for the M seed buyers. It is enough to embed randomly generated fingerprints for the seed buyers such that their pairwise correlation is low. This embedding scheme must fulfill the following conditions:

1. The embedded fingerprint must be a binary sequence spread along the whole content (file). Furthermore, the fingerprint must be separated into pieces which are embedded into different blocks (or fragments) distributed by the P2P software. These (fixed-sized) pieces of the content contain a full segment of the fingerprint. For example, if the P2P software uses 32-kB (kilobyte) fragments, each segment of the fingerprint should be embedded into one of these fragments and the fingerprint extraction method must be robust against fragmentation in 32-kB units, as long as the beginning and the end of the fragments are respected. This process is illustrated in Fig. 2.
   Note that this is not always possible with non-block-based embedding schemes. An example of block-based audio watermarking system which may be used for fingerprinting in this scenario is presented in [20].
2. Even if the versions of the content obtained by different buyers will not be bit-wise identical (the fingerprints embedded into the buyers' copies will differ as a consequence of the P2P distributed download), these versions should be "perceptually" identical, because the distributed content must have the same high quality for all buyers. This means that a standard hash function which produces different hash values even after a single bit change would not be useful in this application. A perceptual hash function [6] for which the same hash value is obtained for different (perceptually identical) versions of the same content would be required if hash values are used for indexing in the P2P distribution software.

If the previous two conditions hold, fingerprinting occurs in an automatic way as contents are obtained by buyers from different sources. No additional overhead for embedding is required. Note that the above automatic fingerprinting requires more than one content source for each buyer to exist: in case of a single source, the fingerprint would be identical for both the source and the buyer. Although some segments of the fingerprint could be modified by running the embedding method in these buyer-to-buyer transfers, this would reduce the simplicity of our proposal. The simplest solution is to enforce at least two parents for each buyer, and this is the choice made in this work.

Even if two children have exactly the same parents, since fragments are picked up randomly from parents, the probability that both children have the same fingerprint is negligible.

We assume that all seed buyers are chosen with equal probability. Thus, on average, all of them contribute with a similar number of fragments to the new buyers. In addition, all subsequent buyers also engage in P2P transfers by being parents of new buyers. Scalability is thus guaranteed by the distribution system even if the number of parents per buyer is small, since the merchant only needs to feed the seed buyers. As new buyers obtain the content, the connections with the seed buyers may become less relevant, since the new buyers become new content sources.

3.3 Co-utility in Parent-Child Relationship

If it can be enforced that there be at least two parents for each buyer, this is the simplest and most effective solution, because, as said above, fingerprinting is automatic in this case. Fortunately, it turns out that it is in the selfish interest of a child buyer to obtain her content from more than one parent, and it is in the selfish interest of a parent buyer to split her content into more than one child buyer:

- If a child obtains her entire content from a single parent then, her fingerprint will be the same as her parent's fingerprint. Then, if the parent happens to illegally redistribute the content, the child risks being unjustly accused of redistribution (see Sect. 5 below). Obtaining the content from several parents is a simple and automatic way to avert that risk.
- If a parent sends her entire content to a single child, her child will inherit the parent's fingerprint. Hence, if the child happens to illegally redistribute the content, the parent risks being unjustly accused of redistribution. Splitting the content among several children is the best option for the parent.

The best strategy to preserve one's own privacy is to act in such a way that someone else's privacy is protected. In game-theoretic terms, the vector of strategies *multiple children, multiple parents* is a Nash equilibrium between parent and child.

The co-utility property (see the first chapter of this book or [10, 12]) ensures that, whenever a child buyer can obtain her content from more than one parent, she will do so; it also ensures that parents will be interested in not passing their entire content to a single child buyer. The latter condition can be easily enforced by the P2P distribution software. For example, when a parent is sending the content through a proxy, it can be enforced that the connection be closed as soon as a given threshold fraction (e.g. 50 or 60%) of the content has been sent. The software can also block any further attempt by the proxy to establish a connection with the same parent for the same content (for some given time window). Each proxy should be forced to choose at least two different parents.

The co-utility analysis given in this section holds for all of our P2P systems described in [19, 21, 22].

3.4 Building Blocks and Notation for Transaction Monitoring and Content Authentication

In order to design protocols for the different steps of the distribution system, the following building blocks are required:

- Public-key cryptography is required in different steps below. Let $E(\cdot, K)$ be the encryption function using the public key K and $D(\cdot, K^s)$ be the decryption function using the private key K^s, required to decrypt a content encrypted using E and K, i.e. $D(E(x, K), K^s) = x$.
- In particular, the transaction monitor uses the following pair of public and private keys: (K_c, K_c^s). Also, each peer node in the network is supposed to have a public key and a private key. A pair of decryption and encryption keys, K_m and K_m^s respectively, for the merchant is also used for the authentication of the fragments.
- For each segment of the fingerprint g_i, a hashing function h produces a 1-bit hash $h(g_i)$. Let h_f be the (ordered) concatenation of the hashes of all segments, called "fingerprint's hash" hereafter. Hence, h_f is constructed as

$$h_f = h(g_1)|h(g_2)| \dots |h(g_l),$$

where l is the number of segments of the fingerprint and "|" stands for the concatenation operator.
- An extraction function exists to obtain the fingerprint from a content. This function receives, as input parameters, the fingerprinted content and a secret extraction key K^e only known by the merchant. This key will be required by the authority to trace an unlawful distribution.

4 The P2P Distribution Protocol

In this section, we present the P2P distribution protocol and how transfers between peer buyers are anonymized. It is based on our ideas in [19, 21, 22].

4.1 System Model

The participants in the proposed fingerprinting system are the following:

- Merchant: he distributes copies of the content legally to the seed buyers. Each fragment of the content contains a different segment of the fingerprint embedded into it. The segments have low pair-wise correlations.
- Seed buyers (B_i for $i = 1, \ldots, M$): they receive fingerprinted copies of the contents from the merchant that are used by the P2P distribution system to bootstrap the system. They can be either real or dummy buyers.
- Other buyers (B_i for $i = M + 1, \ldots, N$, with $M \ll N$): they purchase the content and obtain their fingerprinted copies from the P2P distribution system. The content is assembled from fragments obtained from different "parents". Anonymous connections with peer buyers are provided by means of proxies.
- Proxies: they provide anonymous communication between peer buyers by means of a specific protocol analogous to Chaum's mix networks [5] (see below).
- Transaction monitor: it keeps a transaction register for each purchase carried out for each buyer. This transaction register includes an encrypted version of the embedded fingerprints.
- Tracing authority: in case of illegal re-distribution, it participates in the tracing protocol that is used to identify the illegal re-distributor(s).

4.2 Security Model

The security assumptions of the proposed system are the following:

- The merchant does not need to be trusted either for distribution or to associate a pseudonym with the identity of a buyer. The protocols for distribution and for traitor tracing described below are proven to work even if the merchant is not trusted.
- Buyers are not trusted and protocols are provided to guarantee that (1) they are transferring authenticated fragments of the content and (2) their anonymity can be revoked in case they re-distribute the content illegally.
- The transaction monitor (or any other single party) will not have access to the cleartext of the fingerprints. This prevents that any single party can frame an innocent buyer.
- As privacy is concerned, the transaction monitor is not trusted and it should only have access to pseudonyms, but not to the buyers' real identities.
- The transaction monitor is trusted as the symmetric keys used for encrypting the fragments are concerned. This means that (1) the transaction monitor stores the key provided by each parent buyer and (2) this key can be retrieved only once from its database (in principle by the child buyer). After this retrieval, the transaction monitor blocks the register and eventually removes it (details are provided in the description of Protocol 2 below).

- The transaction monitor returns the true pseudonym corresponding to an illegal re-distributor in the traitor tracing protocol. However, this trust can be replaced by a collection of signatures provided by the proxies.
- The tracing authority is part of the legal system and shall be trusted. It is not expected that the authority participates in any coalition to frame an innocent buyer or break someone's privacy.
- The communication between the merchant and the seed buyers, and between peer buyers within the P2P distribution system, must be anonymous using an onion routing-like approach. The fragments of the content are encrypted using symmetric cryptography.
- Proxies are not trusted and the fragments sent through them shall be encrypted in such a way that only the sender and the recipient have access to their cleartext. Malicious proxies may also try to cheat by reporting false fingerprint segments (or not reporting them at all) to the transaction monitor.
- The fingerprints must be constructed with a long enough number of segments to guarantee that recombination will produce different fingerprints for different buyers due to numerical explosion.
- The hashing functions used in the system are secure and cannot be inverted.
- Public-key cryptography is restricted to the encryption of short binary strings, such as fingerprint segments or hashes. The different parties (merchant, transaction monitor, proxies and buyers) have a pair of public and private keys to be used in different steps of the protocols.
- A single malicious party shall not be able to construct the fingerprinted copy corresponding to any buyer to frame an honest user of the system. In a similar way, a single malicious party shall not be able to link the identity of some buyer to a particular content unless that user is involved in an illegal re-distribution of the content.

Different attacks that may be mounted against the system proposed in this work, regarding both security and privacy, are described below. The following assumptions are made:

- The watermarking method used for embedding and detecting the fingerprint is transparent, robust and secure enough for a fingerprinting application. There are hundreds of watermarking algorithms available to be used with the proposed fingerprinting system and watermarking properties are out of the scope of this work. For example, a robust video watermarking scheme is presented in [27] and a robust audio watermarking scheme is described in [14].
- Collusion attacks are the main topic in most of the research related to fingerprinting. Collusion occurs when several buyers decide to recombine their fingerprinted copies of a given content trying to obtain a new copy in which neither of their fingerprints is detectable. The system suggested in this work inherits the collusion resistance of the method described in [21] and, hence, no further analysis on collusion attacks is required.

Thus, the main attacks that may be performed on the proposed system are related to either the P2P distribution protocol, the traitor-tracing protocol and the P2P network itself. These attacks may be aimed to break either the security or the privacy properties of the system. The attacks to the cryptographic protocols require that one or more of the involved parties are malicious or that a malicious party tries to mimic the behavior of an honest party in order to gain sensitive information that may be used afterwards. As far as security is concerned, there are two main items to be protected:

- **Buyer frameproofness** is related to the possibility that an innocent buyer is accused of illegal re-distribution of the purchased content.
- **Copyright protection** would be broken if any party obtains a copy of the content whose fingerprint is not included in the fingerprints' database of the transaction monitor (and thus can be re-distributed illegally) or the association of that particular fingerprint with the illegal re-distributor cannot be completed.

On the other hand, privacy would be broken if someone could associate a real identity with the purchase of some specific content.

Attack models to frame honest buyers require that another party is able to obtain either the fingerprint of a buyer, or the fingerprinted copy of the content, in such a way that it can be further re-distributed and, finally, the honest buyer is accused of illegal re-distribution. The following types of attacks would be aimed to frame an innocent buyer:

- Random guess: the merchant has access to all fragments of the seed buyers and he may try to recombine the fragments to produce a new copy of the content. If this copy is re-distributed, there is a possibility that it leads to some innocent buyer that could be framed.
- Buyer authentication attacks: an attacker may impersonate a buyer in the system and try to obtain a fingerprinted copy of the content that would be linked to the impersonated buyer.
- Proxy authentication attacks: an attacker may impersonate a proxy in the system and try to obtain fragments of a content from different buyers in order to build a fake "colluded" copy of the content. This fake "collusion" could produce evidence against one ore more honest buyers that served fragments to the fake proxy.
- Man-in-the-middle attacks: an attacker may try to intercept the traffic between a buyer and one or more of her proxies and keep a copy of all the fragments of the content.
- Database authentication attacks: an attacker may try to obtain the fingerprint of a buyer that is stored in the transaction monitor's database.
- Protocol attacks: one or more of the participants in the protocols (proxies, merchant, buyers, transaction monitor or tracing authority) may be malicious and try to obtain the fingerprint or the fingerprinted content linked to a particular honest buyer.

The security of the system against these attacks is discussed in Sect. 7.

4.3 The P2P Distribution Protocol

To bootstrap the system, a few seeds of the fingerprinted content must be produced. The proposed approach is for the merchant to produce a small number M of instances of the content with different pseudo-random binary fingerprints, using some scheme satisfying the conditions described above. These M seeds could be the first buyers of the content who will be the ones contacted by second-generation buyers to obtain further copies of the content. Either the merchant or some trusted authority will keep the association of the first M fingerprints with the identities (or maybe some pseudonym) of the first M buyers. After the system is bootstrapped in this way, all future transactions occur without any further execution of the embedding scheme. Furthermore, all fingerprints from buyer $M + 1$ to the final one are completely anonymous (accessible only if the buyer provides her copy of the content for fingerprint extraction) and do not relate to the buyers' identities. Note that this way of achieving anonymous fingerprinting is much simpler than the anonymous fingerprinting proposals in the literature [1, 3, 8, 25], predicated on some sort of complex cryptographic protocol for *every* transaction. Only the transaction monitor keeps a record of the engaged transactions in case they need to be used in future correlation tests.

We can summarize the P2P distribution protocol as follows.

Protocol 1

1. *For $i := 1$ to M, the merchant generates the i-th seed copy with a random fingerprint embedded in it (the fingerprints of the M copies should have low pairwise correlations).*
2. *For $i := 1$ to M, the merchant forwards the i-th seed copy to the i-th seed buyer. If the seed buyers are genuine rather than dummy buyers, this step can be anonymized as explained below.*
3. *For $i := M + 1$ to N, the i-th buyer obtains her copy of the content by composing fragments obtained from a set S_i of parent nodes such that $S_i \subseteq \{B_1, \ldots, B_{i-1}\}$ and $|S_i| > 1$, where $|\cdot|$ is the cardinality operator and B_j refers to the j-th buyer. This transaction is performed via a proxy (or a set of proxies) and with an anonymous protocol (see below). The proxy registers each transaction at the transaction monitor.*
   Since the same parent may be chosen by different proxies for different fragments, a transaction record for the same parent, child and content may already exist. In that case, no new record would be created. Note, however, that the transaction record will not be complete until all fragments have been obtained by the child buyer from all proxies, since the whole fingerprint hash (which is stored in the transaction record) will not be available until that moment. The transaction record is initially created with temporary information (content, parent, children and date/time) and, when all the fragments have been transferred for a buyer, the whole fingerprint and its hash are also stored in the transaction monitor.

The transaction record stored at the transaction monitor is formed by the following information:

P_i Username (pseudonym) of the buyer B_i.
$H(c)$ Perceptual content hash (used for indexing in the content database).
E_{h_i} Encrypted hash of the buyer's fingerprint.
E_{f_i} Encrypted buyer's fingerprint.
d Transaction date and time (for billing purposes).

Note that the transaction monitor does not store the true identities of the buyers, only pseudonyms. Only the merchant has access to the buyers' database, which relates a given pseudonym to real identity data.

The fingerprint and its hash are not stored as cleartext in the transaction monitor, the buyer's fingerprint is encrypted under the public keys of the parents and the transaction monitor and the hash of the buyer's fingerprint only with the public key of the transaction monitor. Note that having access to the fingerprints' hashes does not allow the transaction monitor to reconstruct any buyer's fingerprint, since a hash function is not invertible, thereby preserving buyer frameproofness.

The storage of an encrypted version of the buyers fingerprints, E_{f_i}, is computed as follows:

- Each fragment of the content shall be transmitted with a fingerprint's segment g_j embedded into it and together with an encrypted version of the segment $E^c_{g_j} \doteq E(g_j, K_c)$, where K_c is the public key of the transaction monitor. A signature can also be included, as detailed in Sect. 7.
- Each proxy selects a set of m contiguous fragments of the content and facilitates the anonymous communication between parents and child for the transmission of those fragments. These m contiguous fragments of the content carry m contiguous segments of the fingerprint embedded into them. The construction of the fingerprint with segments and sets of contiguous segments is shown in Fig. 3.
- The same proxy is required for transferring the m contiguous fragments of the content. In this process, the proxy stores the corresponding encrypted segments $E^c_{g_j}$.
- The proxy concatenates the m contiguous encrypted segments, encrypts the concatenation using the public key of the tracing authority (K_a) and sends the result to the transaction monitor.
- Hence, the transaction monitor stores the following encrypted version of the fingerprint:

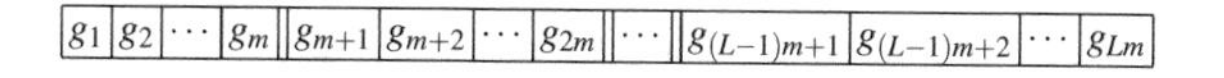

Fig. 3 Fingerprint's segments (g_j) and sets of m contiguous segments

$$E_{f_i} = E\left(E^c_{g_1}|E^c_{g_2}|\dots|E^c_{g_m}, K_a\right)|\dots| \\ E\left(E^c_{g_{(L-1)m+1}}|E^c_{g_{(L-1)m+2}}|\dots|E^c_{g_{Lm}}, K_a\right). \quad (1)$$

Note that (1) no proxy has access to the complete sequence of encrypted segments (since at least two proxies must be chosen by each buyer) and (2) the transaction monitor cannot decrypt E_{f_i} without the private key K_a^s of the authority. In this scheme, the (decrypted) fingerprint's hash h_i would be used in case of collusion of several buyers.

In order to protect the buyers' anonymity, the transfer between buyers must remain anonymous. Otherwise, some buyers (the parents of a child) may collude to generate a replica of the content of another buyer and redistribute it illegally. We refer to the use of symmetric cryptography to encrypt the content in such a way that intermediate routers do not have access to the cleartext of the content. The protocol for anonymous communication is based on using proxies between the parent and the child buyer. The content that is transferred over the proxy is encrypted, again, using symmetric cryptography, but the session key for encrypting the content is shared by parent and child using the transaction monitor as a temporary key database. For a set of fragments, the P2P software runs the following protocol:

Protocol 2

1. *The parent buyer chooses a symmetric (session) key k.*
2. *The parent chooses a pseudorandom binary sequence r to be used as a handle (primary database key) for k. The space for r should be large enough (e.g. 128 bits) to avoid collisions.*
3. *The parent buyer sends* (r, k) *to the transaction monitor, who stores it in a database.*
4. *The parent buyer sends r to the proxy and the proxy forwards r to the child buyer.*
5. *The child buyer sends the handle r to the transaction monitor, who replies with the symmetric key k.*
6. *The transaction monitor blocks the register* (r, k) *for a given period (timer). When the timer expires, the transaction monitor removes the register from the database.*
7. *The parent buyer sends the requested fragments, encrypted with k, to the proxy.*
8. *The proxy forwards all fragments to the child buyer, who can decrypt them using k.*

In this way, only a collusion formed by all the proxies of a child buyer (and possibly the transaction monitor) can replicate the entire fingerprint of the child. If every buyer chooses enough proxies for each content, such a collusion is so unlikely that it can be neglected. Regarding the choice of proxies, possibly the simplest and "most distributed" solution would be that all P2P clients (buyers) can be chosen as proxies by the P2P distribution software. Note that proxies do not have to be buyers of the same content and, thus, this would not break the privacy of buyers. In case that malicious proxies are considered, additional security measures shall be introduced, but this issue is left for future work.

Regarding the payment of content, our protocol does not explicitly consider payment by the buyers to the merchant. Our main focus is on fingerprinted multicast rather than on content sale. In any case, since the transactions are stored in the transaction monitor, a periodic invoice can be issued by the transaction monitor to the merchant such that the merchant can charge the buyers' accounts with the corresponding amounts. Note that such invoice does not need to specify particular contents, since only the total amounts of the downloaded contents of each buyer will be required. This preserves the buyers' privacy with respect to the merchant. It is even possible to establish some prepayment protocol between buyer, transaction monitor and merchant so that the buyer account is charged after each content transfer without disclosing specific contents to the merchant. Another alternative is to protect the access to the P2P platform by means of some subscription account. In any case, payments do not need to be distributed; they can be centralized and simple protocols can be used for them without disclosing which specific contents are being transferred to buyers.

An analysis of the security of this protocol is provided in Sect. 7.

5 Tracing Illegal Redistributors

We now show that the proposed fingerprinting method allows identification of illegal redistributors of fingerprinted contents.

5.1 *Basic Tracing Protocol*

Assuming that the embedding scheme is secure and robust enough so that malicious users cannot easily erase their fingerprints without making the content unusable (this is the standard marking assumption [2]), the following method can be used by a tracing authority to identify the source of an illegally redistributed copy. Notice that the tracing authority is an entity that is independent from both the merchant and the transaction monitor. The basic traitor tracing protocol (when no collusion occurs) begins with the extraction of the fingerprint of the illegally re-distributed copy by the tracing authority. Then, the authority uses the public key of the transaction monitor and its own public key to produce the encrypted fingerprint which can be efficiently searched in the database of the transaction monitor. Once the pseudonym of the illegal re-distributor is available, it can be associated to a real identity.

Protocol 3

1. *The fingerprint f of the illegally redistributed content is extracted by the tracing authority using the extraction method and the extraction key (provided by the merchant).*

2. *The fingerprint's segments* g_j *are encrypted using the public key of the transaction monitor:* $E^c_{g_j} = E(g_j, K_c)$.
3. *The encrypted segments are grouped in sets of m consecutive elements which are encrypted using the public key of the authority, thereby obtaining* E_f *as per (1).*
4. E_f *is (efficiently) searched in the database of the transaction monitor in order to recover the pseudonym of the illegal re-distributor.*
5. *The merchant checks his database of clients and retrieves the identity of the traitor corresponding to the pseudonym obtained in the previous step.*

In the last step of this protocol, to prevent a malicious merchant from cheating and returning the identity of an innocent buyer, the identity must come together with a signed document that certifies the association between the pseudonym and the buyer's identity. All buyers will have to sign such a document (using their private keys) upon registration in the system. These signatures will be verifiable using the buyers' public keys. In this way, (1) there is no need for the merchant to be trusted and (2) non-repudiation from illegal re-distributors with respect to their pseudonym is obtained.

Note that all fingerprints are kept secret except the one that is being traced (f). The advantages of this traitor tracing system are obvious: the cleartext of the fingerprints of honest buyers is never required, and the traitor tracing protocol is based on a a standard database search (not involving innocent buyers).

A security analysis of this protocol is also provided in Sect. 7.

6 Efficiency Aspects

In this section, the *efficiency* of the proposed fingerprinting system is discussed taking into account different points of view. Although simulations could be provided comparing the proposed system with other works of the literature, it must be taken into account that most of the anonymous fingerprinting protocols proposed so far are centralized and, thus, rely on unicast distribution. Simulated experiments to compare unicast and P2P distribution protocols, both as the CPU and communications costs are taken into account, would produce extremely different results, as discussed below. For this reason, a reasoned discussion of this topic (not illustrated with simulations) is provided in this section.

From the merchant's viewpoint, the efficiency of a multimedia content distribution system often refers to scalability. There are two main reasons why the system proposed in this work is scalable. The first one is the use of a P2P distribution system, which reduces the distribution cost significantly on the merchant side: the merchant bootstraps the system by feeding only a few seed buyers, whereas the rest of transactions occur between peer buyers. Among the systems mentioned in the introduction, only [11, 21] propose P2P-based distribution systems. In fact, our system [21, 22] is exactly as efficient as this proposal (which can be found in [19]) from the merchant's point of view. On the other hand, the system [11] requires a multi-party secure protocol between each seed buyer and the merchant to bootstrap the system. Multi-party secure protocols are difficult to implement and require large computational and communication costs.

All the other anonymous fingerprinting systems reported in the introduction [1, 3, 4, 16–18, 23, 26, 28] are centralized. This means that: (1) a unicast distribution is required for each buyer; and (2) an anonymous protocol for embedding the fingerprint and transmitting the fingerprinted file is run between the merchant, each buyer and (possibly) other parties. Obviously, these solutions are much less scalable for the merchant, and involve much higher computational and communicational costs.

Hence, it can be concluded that our systems [19, 21] are the most efficient solutions for a merchant.

As the buyers are concerned, the centralized approaches also involve running expensive protocols, either because of the use of homomorphic encryption of the contents (which require homomorphic decryption in the buyer's side), zero-knowledge proofs, bit commitments or other similar techniques. Only [4] takes the computational costs for the buyers into account and proposes the use of powerful servers for the most expensive operations of the protocols. None of [19, 21] involve expensive decryption of the content (since symmetric encryption is used), zero-knowledge proofs or bit commitments. Although the buyers are required to participate in anonymous communication protocols using symmetric encryption and proxies, these protocols are not particularly demanding in communication or computation terms. For a buyer, the cost associated with obtaining a fingerprinted copy of the content will not be higher than that associated to a centralized approach. However, in P2P distribution systems, the buyers themselves become providers of the contents for other buyers and, thus, their bandwidth will be used in further transactions. Hence, P2P-based systems entail an upload cost for each buyer as long as she keeps the content in her multimedia database.

As far as the proposal of [11] is concerned, it is required that each buyer (except the leaves of the distribution tree) participate in an anonymous fingerprinting protocol with at least another buyer. In such a protocol, a new fingerprint must be embedded for each new buyer, which increases the communication and computational costs for the buyers who become sources of the content. Note that neither [19] nor [21] require that the buyers embed new fingerprints, since the new fingerprints are created through recombination of the segments of existing ones.

In principle, buyers would prefer a centralized distribution system in what regards bandwidth, since once the file has been obtained, no further communications are required. Nevertheless, the fact that the distribution is much "cheaper" from the merchant's point of view implies that the infrastructure (servers) required by the merchant is much simpler. The reduced costs as compared with centralized distribution and fingerprinting will certainly have some effect in the price of the contents. If buyers enjoy cheaper prices, this would compensate them for the use of their bandwidth required by the P2P distribution protocol.

Another disadvantage of [19], as far as efficiency is concerned, is the need of an anonymous communication protocol, analogous to Chaum's mix networks [5], between the peer buyers. This protocol increases the number of messages in the network as a consequence of the use of proxies. In any case, this is a general disadvantage of any anonymous communication system, since a direct connection between sender and recipient must be avoided.

In short, our systems [19, 21] can be considered efficient from both the merchant's and the buyers' points of view, but the buyers are requested to become providers of the content for other buyers (which is a standard drawback of any P2P distribution system).

7 Security Analysis

In this section, we first specify the security assumptions of our scheme [19]. We then analyze more complex collusion attacks.

As detailed in Sect. 4.2, attacks to the system may be classified as authentication/impersonation attacks, man-in-the-middle attacks and protocol attacks. Authentication/impersonation attacks should be overcome by using existing secure authentication protocols and are out of the scope of this work. As man-in-the-middle attacks are concerned, there is no possibility of intercepting and decrypting the messages between a buyer and a proxy, since communications with the transaction monitor and the child buyer should also be attacked in order to obtain the session key used for encrypting the content. If the communication between the child buyer and the transaction monitor (Step 5 of Protocol 2) are strongly authenticated (e.g. using a Public Key Infrastructure), the possibility of a successful man-in-the-middle attack can be neglected.

7.1 Formal Analysis of the Proposed Protocols

First of all, the security and privacy properties of Protocols 2 and 3 is analyzed by means of two theorems (and their corresponding proofs).

Theorem 6.1 *In Protocol 2, a malicious proxy trying to decrypt the fragments of the content would be detected.*

Proof If a malicious proxy tries to obtain the session key k by sending r to the transaction monitor there are two possibilities:

1. If the child buyer has already retrieved k from the database by sending the handle r to the transaction monitor, the register containing k would be either blocked or removed. Note that the transaction monitor is assumed to be honest for the management of the symmetric keys (see Sect. 7).
2. If the child buyer has not retrieved k from the transaction monitor, the proxy will obtain it, but the child buyer will find the corresponding register either blocked or removed. Then, the malicious behavior of the proxy can be reported to the authorities and the transaction monitor and the child buyer have enough information (such as pseudonyms and IP addresses) to identify the misbehaving proxy.

Again, the assumption of honest behavior for the management of symmetric keys (Sect. 7) applies.

Hence, a malicious proxy trying to obtain k from r would be detected, since the register would be blocked either to the proxy or to the child buyer, raising an investigation. This completes the proof.

Theorem 6.2 *By applying Protocol 3, an illegal re-distributor can be traced efficiently using a standard database search in the transaction monitor and it is not required to decrypt any of the fingerprints recorded by the transaction monitor. The output of the tracing protocol is the identity of at least one illegal re-distributor.*

Proof If no collusion occurs, the fingerprint f would be first extracted by the tracing authority, which is trusted (Sect. 7). Then the tracing authority would compute $E^c_{g_j} = E(g_j, K_c)$ for each segment (using the public key of the transaction monitor), and finally obtain E_f after grouping the segments in sets of m consecutive elements and encrypting these groups with its public key K_a. After that, the transaction monitor, which is also trusted for transaction database search (Sect. 7), would output the pseudonym of the illegal re-distributor. The pseudonym can be linked to the real identity by the merchant, who provides also a signed document that associates the real identity and the pseudonym. This completes the proof.

In case of collusion of several buyers, the extracted fingerprint would not be a valid codeword of the anti-collusion code used in the scheme. Then, the system described in Sect. 2 would be used: the encrypted hash $E_{h_f} = E(h_f, K_c)$ would be searched instead of the encrypted fingerprint, where h_f denotes the hash obtained applying the hash function to the traced fingerprint f. Thus, Protocol 3 would be used with the hash of the fingerprint instead of the fingerprint itself. As described in Sect. 3.1, with a large enough hash space, hash collisions would be almost negligible and a traitor would still be identified in the vast majority of the cases.

The requirement that the transaction monitor is trusted and returns the pseudonym of the buyer associated with the traced fingerprint (and not a different pseudonym) can be relaxed if a signature of the encrypted sets of segments of the fingerprint is provided by the proxies. These signatures can be verified using the public keys of the proxies. In that case, both the signatures and the pseudonyms of the proxies shall also be included in the registers of the transaction database to facilitate the verification of these signatures when required.

7.2 Collusion Attacks on the Protocols

Fingerprinting schemes must provide some degree of collusion resistance in order to be able to trace forged copies created by advanced attackers. In this section, we discuss possible collusion attacks on the proposed protocols. Hence, our scheme can be made as resistant against collusion as any of the existing anti-collusion techniques of the literature.

Under the usual marking assumption, error correcting codes are a typical solution to detect collusions [9]. Other approaches are based on more recent techniques, such as Tardos codes [29] or even newer codes based on them, like [24]. In the latter case, the marking assumption can be relaxed to a δ-marking assumption [24].

7.2.1 Buyer Frameproofness

Buyer frameproofness relates to the case of a malicious merchant trying to frame an honest buyer by accusing her of being the source of an illegally redistributed content. In this scheme, the merchant is not able to produce any buyer's fingerprint by random guess due to the numerical explosion of the fingerprint space, even with a reduced number of seed buyers. On the other hand, the transaction monitor has access only to the hashes of the fingerprints (not the fingerprints themselves without the private key of the authority). Since the hash function is not invertible, it is not possible for the monitor (even in coalition with the merchant) to reconstruct any buyer's fingerprint. Possible collusions to disclose the specific fingerprint of an innocent buyer are the following:

1. The tracing authority and the transaction monitor.
2. All the proxies (for a transfer) and the transaction monitor.
3. All the proxies (for a transfer) and the merchant.

In the first case, the authority and the transaction monitor may use their private keys to obtain the cleartext of all the fingerprints. However, this possibility can be neglected since at least the authority must be trusted (as described in Sect. 4.2). In the second case, all the segments of the fingerprint could be decrypted using the private key of the transaction monitor, since the malicious proxies would not encrypt them with the public key of the authority. Also, the transaction monitor could collude with the proxies and use the session keys k to decrypt the fragments. Both possibilities would involve at least three malicious parties: all the proxies (two at least per each purchase) and the transaction monitor. In the third case, even if the transaction monitor does not provide her private key, a brute force attack segment by segment would be possible to reconstruct a buyer's fingerprint, because the number of different segments is small for each fragment (equal to M). Again, at least three malicious parties would be required: two (or more) proxies plus the merchant.

Hence, the minimum coalition required to frame an innocent buyer is formed by three malicious parties (or two if one of them is the authority). Note that a coalition of the transaction monitor and the merchant is not enough to obtain the cleartext of any fingerprint. As the proxies encrypt a set of m consecutive segments, and there are M possible values for each segment, the total number of combinations per set of consecutive segments is M^m. This avoids a brute force attack if m is reasonably large. For example, if $M = 10$ and $m = 32$, there would be 10^{32} possible combinations for each set of consecutive segments, what would be enough for security against a brute force attack. If the segments were encrypted one by one (or grouped with a small value of m), the system would be vulnerable against a brute force attack for a collusion of the merchant and the transaction monitor.

7.2.2 Copyright Protection

In order to ensure copyright protection, it is essential that the fingerprint embedded in each buyers' copy of the content and its encrypted version recorded by the transaction monitor are identical. If there is a way to cheat in the recorded fingerprint, the corresponding buyer would be able to re-distribute her copy illegally without any chance of being detected. As already remarked in Sect. 4, the content fragments are signed by the merchant from origin. The same approach can be used here for each encrypted segment of the fingerprint, making it impossible for a proxy to cheat about the fingerprint. The authority and the merchant could verify randomly, with some probability, the signatures of the set of contiguous segments reported by a proxy. If the signature was not verified, the proxy would be accused of forgery. Note that the fingerprints would still be protected since (1) only some sets of contiguous segments would be verified (not the whole fingerprint) and (2) those segments would still be encrypted with the transaction monitor's public key.

However, a proxy may still try to get alternative fragments for the same position of the content by requesting them from different parents. That possibility would allow the proxy to cheat about the true fingerprint of the child buyer, since several correctly signed fragments would be available for him for the same content. This behavior can be avoided in several ways. For example, temporary records can be created in the transaction monitor by the parents to detect if a proxy tries to obtain two alternative fragments for the same content.

7.2.3 Buyers' Privacy

The identity of a buyer who has purchased a specific content could be revealed by a coalition of two parties: one of the proxies chosen by the buyer and the merchant (who can link her pseudonym to a real identity) or, similarly, the transaction monitor and the merchant. Better privacy could be achieved if, for example, the pseudonyms were encrypted by the proxies using the public key of the tracing authority. In that case, a coalition of the merchant and the transaction monitor would not be enough to break a buyer's privacy, but a coalition of a proxy and the merchant would still be enough. However, the merchant should not be interested, in principle, to break her customers' privacy, since privacy would be one of the clear advantages of the proposed distribution system.

Another threat to privacy is the fact that all anonymous communications between each child and each parent occur through a unique proxy. This means that this proxy has access to different pseudonyms (the parents' and the child's). This can be easily circumvented if more proxies are used in Protocol 2 between child and parent. With two proxies, each of them would know only the pseudonym of one of the parties (although they could still collide). With three or more proxies, only two of them would have access to different pseudonyms (either the parents' or the child's). Of course, increasing the number of proxies in each transfer would affect the efficiency of the system, since more communication burden would be required.

8 Simulation Results

This section presents a set of simulated experiments to illustrate the properties of the proposed system: buyer privacy, robustness against non-collaborative buyers and collusion resistance.

8.1 Buyer Privacy

In all simulations presented in this section, the recombined fingerprints were 4096-bit sequences divided into 128 segments of 32 bits each. The first simulation to confirm the results presented in Sect. 7.2.3 consisted of producing different generations of buyers using an exponential growth approach and checking the average number of correlation tests required to identify the buyers. The number of seed buyers was taken to be $M = 10$ and each buyer could have between two and four parents which were chosen at random from all the previous generations (not only the immediately previous one). This means that the average number of parents per non-seed buyers was $n = 3$. The simulations shown in Table 1 were carried out, and a comparison of the average number of correlation tests with the expected fraction introduced in Sect. 7.2.3 is shown in Fig. 4.

The results in Table 1 show a single simulation and the average 100 simulations with 100 different seeds in the pseudo-random number generator in order to reduce the bias of the results. It can be seen that no significant differences appeared between 1 and 100 simulations. The last column represents the average percentage of buyers requiring backtracking in the 100 simulations. Not surprisingly, as the network (graph) became larger, more buyers required backtracking, but the percentage was always small.

Figure 4 shows intervals for the average fraction of non-seed buyers affected by correlation tests as the number of generations grew. For each number of generations, the corresponding vertical solid line represents an interval with the up triangle showing the maximum fraction in 100 simulations, the down triangle showing the minimum fraction and the circle showing the average fraction; these average fractions correspond to the percentages given in Table 1 for 100 simulations. As discussed in Sect. 7.2.3, the theoretical expected maximum fraction of tested non-seed buyers (dashed line) can be exceeded if the number of generations is small, due to the effect of some parents having more than the average number of children. This situation is compensated as more generations are produced and the simulated fraction goes

Table 1 Average number and percentage of correlation tests on non-seed buyers in an exponentially growing population

Generation	Population	Average correlation tests		Backtracking (100 sim. %)
		1 simulation	100 simulations	
$k = 2$	$N = 20$	3.40 (34.0%)	3.71 (37.1%)	0
$k = 3$	$N = 40$	6.93 (23.1%)	7.29 (24.3%)	0
$k = 4$	$N = 80$	12.26 (17.5%)	11.69 (16.7%)	0.6
$k = 5$	$N = 160$	18.99 (12.7%)	17.05 (11.4%)	1.2
$k = 6$	$N = 320$	24.31 (7.8%)	23.76 (7.7%)	2.7

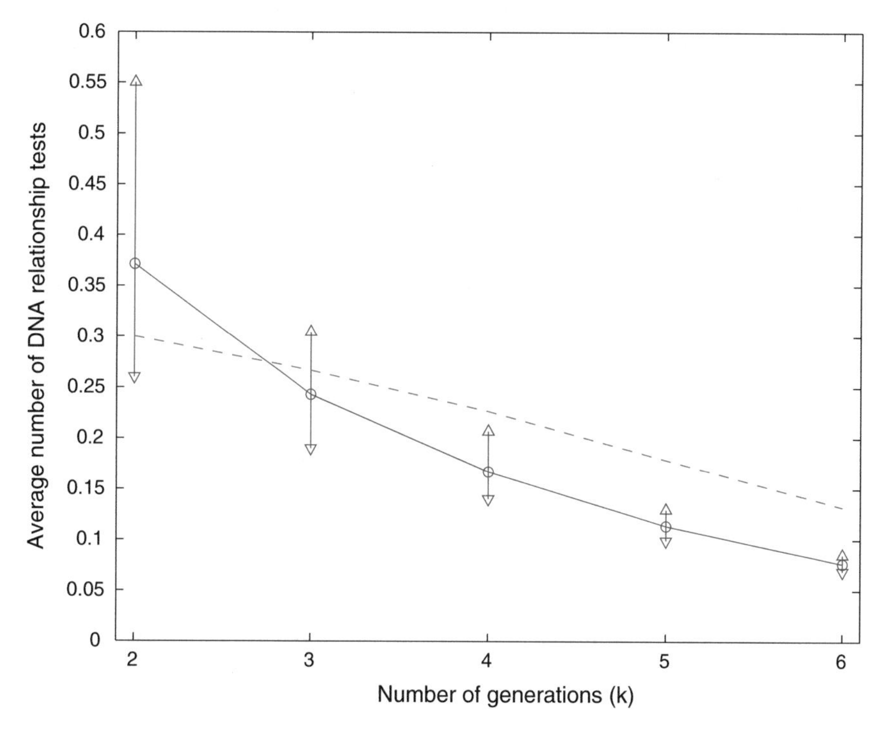

Fig. 4 Average fraction of non-seed buyers affected by correlation tests in an exponentially growing population: simulation results (*solid*) and theoretical expected maximum (*dashed*). *Vertical solid lines* are max-min intervals

below the theoretical value already for $k > 3$, although the interval for $k = 3$ shows that, for that number of generations, some simulations still yielded fractions above the theoretical value. In any case, as predicted in Sect. 7.2.3, the fraction of non-seed buyers affected by *one* correlation test decreased to zero as the number of generations grew: the more buyers involved, the higher the probability that a buyer did not need to surrender her privacy in one particular correlation test. However, as the population grows, the number of illegal redistributions may also increase and more correlation tests may be needed to investigate them; as the number of required correlation tests increases, the probability that a non-seed buyer is affected by them (and therefore loses her privacy) also increases.

One should notice that correlations would be quite small if the number of parents providing a significant number of fragments to each child is too large. In particular, it is not advisable that the number of parents be close to the number M of seed buyers. Intuitively, if the number of parents is close to M, all correlations will be very small and similar, hence leading to more wrong choices in the tracing algorithm and, thus, more backtracking and more tests. We have confirmed this issue in additional simulation results that are omitted here for brevity. As a rule of thumb, the distribution software should limit the number of parents (through the number of proxies) such that it is not too close to M. Since specific software is to be used for the P2P distribution, constraints can be enforced on the number of allowed proxies/parents for each content transfer. In fact, if many parents are allowed for a buyer, the correlations between parents and children will be relatively low, irrespective of the number M of seed buyers, which would lead to more backtracking. It is thus advisable to keep the number of allowed parents limited even if the number of seed buyers is relatively large.

It may appear that the percentage of buyers involved in correlation tests in the course of an investigation decreases to zero because of the exponential increase in population occurring at each generation. However, this is not the case. The decrease of this ratio of tested buyers depends on the population and not on the particular way it grows. To illustrate this process, the following simulations were performed with a population growing linearly at each generation:

1. The first generation was, again, formed by the $M = 10$ seed buyers who obtain their fingerprinted contents from the merchant.
2. At each new generation, $M = 10$ new buyers obtained their contents from a variable number of parents between two and four (and thus, the average number of parents was, again, $n = 3$).
3. With this scenario, the population N increased linearly with the number of generations: there were $N = kM$ buyers after the k-th generation.

Table 2 illustrates this issue. It can be seen that the fraction of tested buyers decreased with the number of generations. In this case, the decrease was linear and not exponential, since the population increased linearly with k. This is also illustrated in Fig. 5 by means of interval plots. The seeds of the pseudo-random number generator were adjusted such that the results for two generations ($N = 20$) were the same as those presented in Table 1 for the exponential growth.

We present also simulation results comparing the linear and exponential growths scenarios *for the same population*. The results are shown in Table 3 and Fig. 6. It can be seen that, when populations are of the same size, the results are almost identical irrespective of the number of generations and the growth model (exponential or linear). Again, the seeds of the pseudo-random number generator were adjusted so that the results for two generations ($N = 20$) were the same for both growth models.

Table 2 Average number and percentage of correlation tests on non-seed buyers in a linearly growing population

Generations	Population	Average correlation tests		Backtracking (100 sim. %)
		1 simulation	100 simulations	
$k = 2$	$N = 20$	3.40 (34.0%)	3.71 (37.1%)	0
$k = 3$	$N = 30$	5.40 (27.0%)	5.54 (27.7%)	0
$k = 4$	$N = 40$	6.20 (20.7%)	6.76 (22.5%)	0.17
$k = 5$	$N = 50$	7.45 (18.6%)	8.15 (20.4%)	0.45
$k = 6$	$N = 60$	8.20 (16.4%)	8.95 (17.9%)	0.42

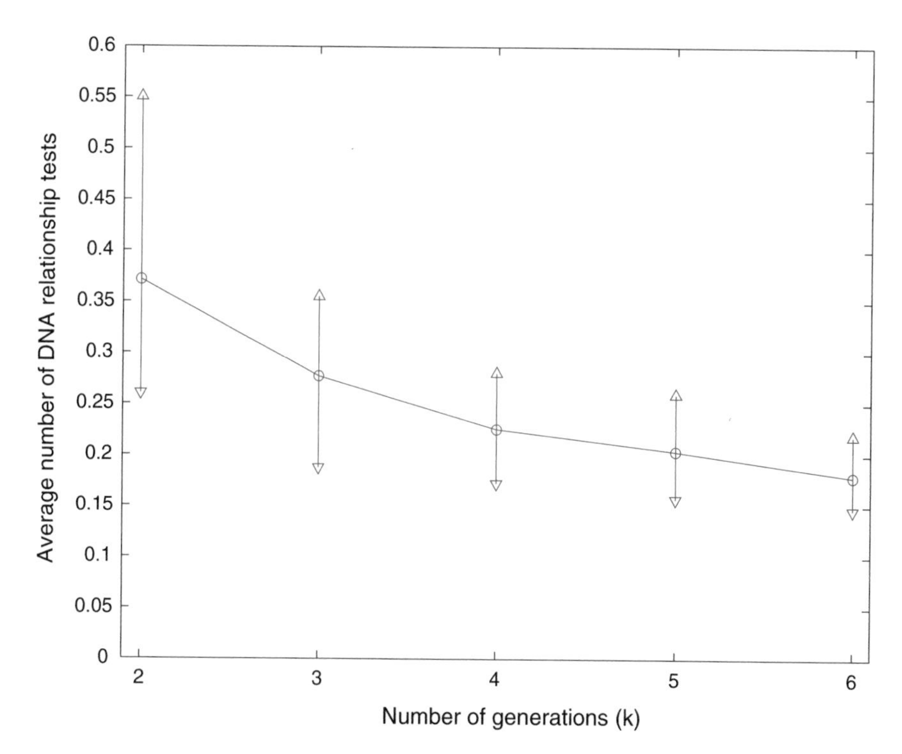

Fig. 5 Average fraction of non-seed buyers affected by correlation tests in a linearly growing population (simulation results). *Vertical solid lines* are max-min intervals

8.2 Collusion Resistance

In this section, experiments conducted with the anti-collusion version of the scheme suggested in Sect. 7.2 are presented. This simulation is a proof of concept. In a practical implementation, other codes and parameters should possibly be used. However, this implementation shows that the method suggested to fight collusion is more than a theoretical possibility.

Table 3 Average number and percentage of correlation tests on non-seed buyers: comparison between exponential and linear growth for the same population

Population	Exponential growth		Linear growth	
	Generations	Average tests (100 simul.)	Generations	Average tests (100 simul.)
$N = 20$	$k = 2$	3.71 (37.1%)	$k = 2$	3.71 (37.1%)
$N = 40$	$k = 3$	7.29 (24.3%)	$k = 4$	6.90 (23.0%)
$N = 80$	$k = 4$	11.69 (16.7%)	$k = 8$	10.62 (15.2%)
$N = 160$	$k = 5$	17.05 (11.4%)	$k = 16$	15.43 (10.3%)
$N = 320$	$k = 6$	23.76 (7.7%)	$k = 32$	22.23 (7.2%)

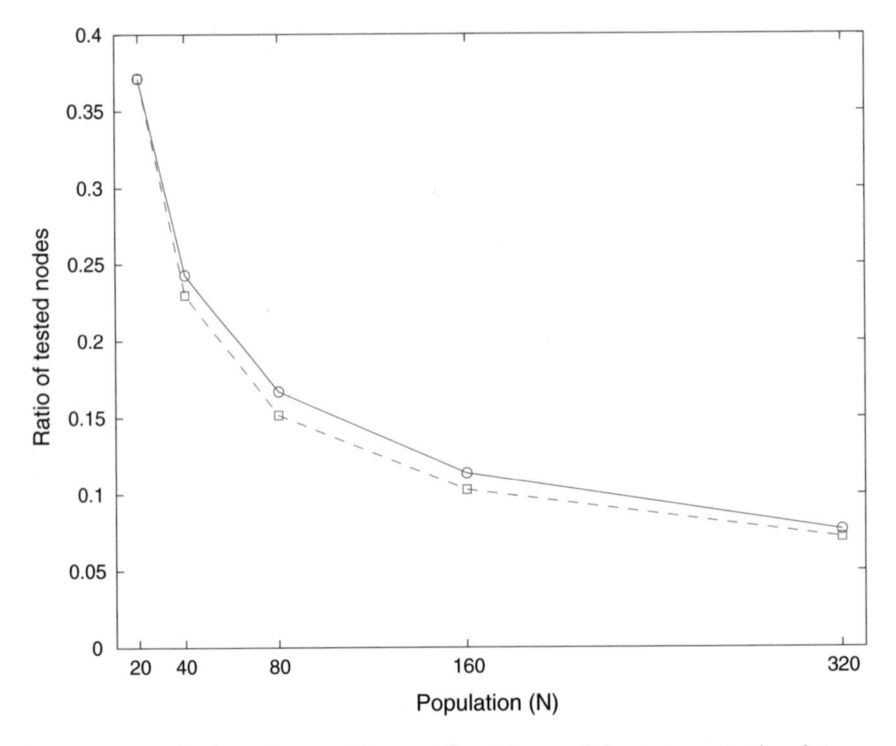

Fig. 6 Average fraction of non-seed buyers affected by correlation tests: comparison between exponential growth (*circle*, *solid*) versus linear growth (*square*, *dashed*) for the same population. Abscissa is population size

The details of the implementation are as follows:

- A dual Hamming code $DH(31, 5)$ was used to encode the segments. $2^5 = 32$ values were thus possible for each segment. Each segment had 5 bits of data and 26 redundancy bits. This code can be used to detect collusions of two buyers.

- A dual Hamming code $DH(1023, 10)$, which also detects collusions of two buyers, was used to encode the hash of the fingerprint. Hence, $2^{10} = 1024$ different hashes existed, with 10 bits of data and 1013 bits of redundancy. This number of hashes would not be enough for a real implementation of the method, but it sufficed for this proof of concept.
- With these choices, the fingerprints were formed based on 1023 segments, each of which consisting of 31 bits. Hence, the fingerprints were $1023 \cdot 31 = 31{,}713$-bit long. The multimedia content had to be split into 1023 fragments, carrying each 31 embedded bits. Possibly, a better choice for a practical implementation with error-correcting codes would be Reed-Solomon (RS) codes instead of dual Hamming codes. In that case, the segments would represent symbols of the code (segment-level code) and the hash of the fingerprint could be an RS codeword (hash-level code). Note that high-capacity robust watermarking schemes exist for embedding that amount of information. For example, the method proposed in [13] allows embedding up to 11,000 bits in a second of audio.
- 10 seed buyers were generated ($M = 10$). The hash of each segment was computed by simply selecting the third data bit of each gene. This is not a sophisticated hash and is obviously quite insecure, but it sufficed for simulation purposes. More advanced hashing techniques would be required in practice.
- For each segment, exactly five seed buyers had a '0' hash and the other five had a '1' hash. As pointed out in Sect. 7.2, this maximized the chances that a proxy could find parents with segments having the hash bit values required to build any hash-level anti-collusion codeword.
- An exponential increase of the population was assumed: 6 generations were created, resulting in a total population of $10 \cdot 2^5 = 320$ buyers (including the seed buyers).
- When non-seed buyers downloaded the content, the first 10 segments were chosen randomly between two and four parents. This yielded the 10 data bits of the hash of the fingerprint. The remaining 1013 bits of the hash had to be such that a codeword of the $DH(1023, 10)$ code was obtained. This was achieved by requesting fragments with segments that carry the appropriate hash bit to the current set of parents. If no parent with the required bit was found, the proxy looked for a new parent with an appropriate segment. This new parent was included in the set of parents of that buyer and was considered as a potential parent for the remaining fragments (segments and hash bits).

After generating a random population with these settings, the actual number of parents per (non-seed) buyer ranged from 4 to 11, with an average of $n = 9.09$ parents per buyer. Hence the privacy results could not be directly compared with those of the previous sections (for which the average number of parents per buyer was around $n = 3$). Note that the number of parents needs to be increased in the collusion-resistant case compared to the basic case because the bits of the fingerprint's hash cannot be chosen in a completely free manner. A valid codeword must be constructed for the fingerprint's hash. Thus, extra parents are selected during the P2P download if none of the already chosen parents provides the required hash bit for a given fragment.

For example, it is highly unlikely that the new fingerprint hash can be constructed completely with only two parents since the collusion-resistant code may require a specific hash bit ('0' or '1') for a particular position, and this hash bit may not be available in the corresponding fragment of these two parents. In fact, the minimum number of parents obtained during this simulation to construct a valid codeword for the fingerprint's hash was four.

With these settings, 200 random bit collusions were generated. For each collusion, a new fingerprint was created by choosing randomly a new fingerprint's bit when the bits of the colluders differed. Hence, after the collusion, the obtained forged copy had a non-codeword embedded into it, both at the segment level and at the hash level. This is the standard marking assumption. For each forged copy, the advanced tracing system described in Sect. 7.2 was applied, by decoding the segments and the fingerprint's hash using the $DH(31, 5)$ and $DH(1023, 10)$ codes, respectively. Note that, with this approach, the colluders themselves did not need to participate in the search. When a buyer was selected as the most likely ancestor of the colluder, the hashes of her children were examined. If a match occurred for the hash of the fingerprint, the corresponding child buyer was the traitor. In case that hash collisions are allowed, some additional investigations would be required to guarantee that the selected buyer is a colluder, but this simplified scenario did not require further tests. After these 200 experiments, the average number of tests to find the colluder (with neither false positives nor false negatives) was 47.77 or 14.8% of non-seed buyers. This is much below the theoretical maximum expected value for $M = 10$, $n = 9.09$ and $k = 6$ as $R = 0.400$ or 40.0% of non-seed nodes. Thus, even in case of collusion, the number of non-seed nodes involved in correlation tests decreases to zero as the population grows.

9 Conclusions

We have presented a recombination fingerprinting P2P scheme. We show that, through co-utility, the proposed solution enforces automatic fingerprinting because merchants and buyers are selfishly interested in distributing or obtaining contents, respectively, from more than one buyer or merchant. The proposed scheme allows the merchant to trace unlawful redistributors of the P2P distributed content. The merchant knows at most the fingerprinted copies of the seed buyers, but not the fingerprinted copies of non-seed buyers (the vast majority). Hence, the merchant does not know the identities of non-seed buyers. Whenever an illegal redistribution needs to be traced, only a small fraction of honest users must surrender their privacy by providing their fingerprinted copies (quasi-privacy). Our scheme also offers collusion resistance against dishonest buyers trying to create a forged copy without any of their fingerprints. Finally, a malicious merchant is most likely to fail using the fingerprinted copies of seed buyers to try to frame an honest non-seed buyer (buyer frameproofness).

Acknowledgements Funding by the Templeton World Charity Foundation (grant TWCF0095/AB60 "CO-UTILITY") is gratefully acknowledged. Also, partial support to this work has been received from the Government of Catalonia (ICREA Acadèmia Prize to J. Domingo-Ferrer and grant 2014 SGR 537), the Spanish Government (projects TIN2014-57364-C2-1-R "SmartGlacis", TIN2015-70054-REDC and TIN2016-80250-R "Sec-MCloud") and the European Commission (projects H2020-644024 "CLARUS" and H2020-700540 "CANVAS"). The authors are with the UNESCO Chair in Data Privacy, but the views in this work are the authors' own and are not necessarily shared by UNESCO or any of the funding bodies.

References

1. Bo, Y., Piyuan, L., Wenzheng, Z.: An efficient anonymous fingerprinting protocol. In: Computational Intelligence and Security, LNCS, vol. 4456, pp. 824–832. Springer, Berlin (2007)
2. Boneh, D., Shaw, J.: Collusion-secure fingerprinting for digital data. In: Advances in Cryptology, LNCS, vol. 963, pp. 452–465. Springer, Berlin (1995)
3. Camenisch, J.: Efficient anonymous fingerprinting with group signatures. In: Asiacrypt, LNCS, vol. 1976, pp. 415–428. Springer, Berlin (2000)
4. Chang, C.-C., Tsai, H.-C., Hsieh, Y.-P.: An efficient and fair buyer-seller fingerprinting scheme for large scale networks. Comput. Secur. **29**(2), 269–277 (2010)
5. Chaum, D.L.: Untraceable electronic mail, return addresses, and digital pseudonyms. Commun. ACM **24**(2), 84–90 (1981)
6. Cox, I.J., Miller, M.L., Bloom, J.A., Fridrich, J., Kalker, T.: Digital Watermarking and Steganography. Morgan Kaufmann, Burlington (2008)
7. Domingo-Ferrer, J.: Anonymous fingerprinting of electronic information with automatic identification of redistributors. Electron. Lett. **34**(13), 1303–1304 (1998)
8. Domingo-Ferrer, J.: Anonymous fingerprinting based on committed oblivious transfer. In: Public Key Cryptography-PKC 1999, LNCS, vol. 1560, pp. 43–52. Springer, Berlin (1999)
9. Domingo-Ferrer, J., Herrera-Joancomartí, J.: Short collusion-secure fingerprints based on dual binary Hamming codes. Electron. Lett. **36**(20), 1697–1699 (2000)
10. Domingo-Ferrer, J., Martínez, S., Sánchez, D., Soria-Comas, J.: Co-utility: self-enforcing protocols for the mutual benefit of participants. Eng. Appl. Artif. Intell. **59**, 148–158 (2017)
11. Domingo-Ferrer, J., Megías, D.: Distributed multicast of fingerprinted content based on a rational peer-to-peer community. Comput. Commun. **36**, 542–550 (2013)
12. Domingo-Ferrer, J., Sánchez, D., Soria-Comas, J.: Co-utility: self-enforcing collaborative protocols with mutual help. Prog. Artif. Intell. **5**(2), 105–110 (2016)
13. Fallahpour, M., Megías, D.: High capacity audio watermarking using the high frequency band of the wavelet domain. Multimed. Tools Appl. **52**(2), 485–498 (2011)
14. Fallahpour, M., Megías, D.: Secure logarithmic audio watermarking scheme based on the human auditory system. Multimed. Syst. **20**, 155–164 (2014)
15. Goldberg, D.: Genetic Algorithms in Search. Optimization and Machine Learning. Addison-Wesley, Boston (1989)
16. Katzenbeisser, S., Lemma, A., Celik, M., van der Veen, M., Maas, M.: A buyer-seller watermarking protocol based on secure embedding. IEEE Trans. Inf. Forensics Secur. **3**(4), 783–786 (2008)
17. Kuribayashi, M.: On the implementation of spread spectrum fingerprinting in asymmetric cryptographic protocol. EURASIP J. Inf. Secur. **2010**, art 694797 (2010)
18. Lei, C.-L., Yu, P.-L., Tsai, P.-L., Chan, M.-H.: An efficient and anonymous buyer-seller watermarking protocol. IEEE Trans. Image Process. **13**(12), 1618–1626 (2004)
19. Megías, D.: Improved privacy-preserving P2P multimedia distribution based on recombined fingerprints. IEEE Trans. Dependable Secur. Comput. **12**(2), 179–189 (2015)

20. Megías, D., Serra-Ruiz, J., Fallahpour, M.: Efficient self-synchronised blind audio watermarking system based on time domain and FFT amplitude modification. Signal Process. **90**(12), 3078–3092 (2010)
21. Megías, D., Domingo-Ferrer, J.: DNA-inspired anonymous fingerprinting for efficient peer-to-peer content distribution. In: Proceedings of the IEEE Congress on Evolutionary Computation (CEC 2013), pp. 2376–2383 (2013)
22. Megías, D., Domingo-Ferrer, J.: Privacy-aware peer-to-peer content distribution using automatically recombined fingerprints. Multimed. Syst. **20**(2), 105–125 (2014)
23. Memon, N., Wong, P.-W.: A buyer-seller watermarking protocol. IEEE Trans. Image Process. **10**(4), 643–649 (2001)
24. Nuida, K., Fujitsu, S., Hagiwara, M., Kitagawa, T., Watanabe, H, Ogawa, K., Imai, H.: An improvement of Tardos's collusion-secure fingerprinting codes with very short lengths. In: Proceedings of the 17th International Conference on Applied Algebra, Algebraic Algorithms and Error-correcting Codes (AAECC'07), pp. 80–89. Springer, Berlin (2007)
25. Pfitzmann, B., Waidner, M.: Anonymous fingerprinting. In: Advances in Cryptology-EUROCRYPT'96, LNCS, vol. 1233, pp. 88–102. Springer, Berlin (1997)
26. Pfitzmann, B., Sadeghi, A.-R.: Coin-based anonymous fingerprinting. In: Advances in Cryptology-EUROCRYPT'99, LNCS, vol. 1592, pp. 150–164. Springer, Berlin (1999)
27. Preda, R.O., Vizireanu, D.N.: Robust wavelet-based video watermarking scheme for copyright protection using the human visual system. J. Electron. Imaging **20**, 13022–130030 (2011)
28. Prins, J.P., Erkin, Z., Lagendijk, R.L.: Anonymous fingerprinting with robust QIM watermarking techniques. EURASIP J. Inf. Secur. **2007**, art 20 (2007)
29. Tardos, G.: Optimal probabilistic fingerprint codes. In: Proceedings of the 35th Annual ACM Symposium on Theory of Computing (STOC '03), pp. 116–125. ACM, New York (2003)

Co-utile Ridesharing

David Sánchez, Sergio Martínez and Josep Domingo-Ferrer

Abstract Ridesharing has the potential to bring a wealth of benefits both to the actors directly involved in the shared trip (e.g., shared travel costs or access to high-occupancy vehicle facilities) and also to the society in general (e.g., reduced traffic congestion and CO_2 emissions). However, even though ridesharing is based on a win-win collaboration and modern mobile communication technologies have significantly eased discovering and managing ride matches, the adoption of ridesharing has paradoxically decreased during the last years. In this respect, recent studies have highlighted how privacy concerns and the lack of trust among peers are crucial issues that hamper the success of ridesharing. In this chapter, we tackle both of these issues by means of (i) a fully decentralized P2P ridesharing management network that avoids centralized ride-matching agencies (and hence private data compilation by such agencies); and (ii) an also decentralized reputation management protocol that brings trust among peers, even when they have not previously interacted. Our proposal rests on co-utility, which ensures that rational (even purely selfish) peers will find no incentives to deviate from the prescribed protocols. We have tested our system by using data gathered from real mobility traces of cabs in the San Francisco Bay area, and according to several metrics that quantify the degree of adoption of ridesharing and the ensuing individual and societal benefits.

D. Sánchez (✉) · S. Martínez · J. Domingo-Ferrer
UNESCO Chair in Data Privacy, Department of Computer Science and Mathematics,
Universitat Rovira i Virgili, Av. Països Catalans 26, 43007 Tarragona, Catalonia, Spain
e-mail: david.sanchez@urv.cat

S. Martínez
e-mail: sergio.martinezl@urv.cat

J. Domingo-Ferrer
e-mail: josep.domingo@urv.cat

J. Domingo-Ferrer and D. Sánchez (eds.), *Co-utility*, Studies in Systems,
Decision and Control 110, DOI 10.1007/978-3-319-60234-9_7

1 Introduction

Ridesharing is a mode of transportation in which several travelers share a vehicle (typically a private car) for a trip and split the trip costs. In this manner, they can enjoy the convenience and speed of private car rides without paying much more than if using public transportation [13]. Moreover, ridesharing is often seen as a promising means to reduce CO_2 emissions [5] and use finite oil supplies in a wiser way. For end users (drivers and passengers), ridesharing mainly saves travel costs, but it may also reduce travel time in case high-occupancy vehicle (HOV) facilities are available for vehicles carrying two or more people (special lanes, toll booths or parking spaces) and, if widely adopted, by mitigating traffic congestion [6, 15]. In fact, the benefits of congestion mitigation can be very significant and extend far beyond the ridesharers themselves: the annual cost of traffic congestion in the US in terms of wasted time and fuel was estimated at $78 billion in 2007 [22].

Ridesharing naturally defines a win-win scenario whereby the involved parties maximize their outcomes as a result of their collaboration. This mutually beneficial collaboration has been formalized and termed *co-utility* (see [11, 12] and the first chapter of this book). Ridesharing is co-utile because, in terms of cost and time, it is the best option even for purely selfish agents [21].

Ridesharing has been used on a regular basis since the 1970s by means of *carpools*. Carpooling is usually understood as an organized and regular ridesharing service (e.g., employees taking turns to drive each other to work). With the advent of the Internet and the enormous adoption of mobile communication technologies, modern ridesharing is also characterized by a more dynamic (or even real-time) scheduling of rides [6]. By dynamic ridesharing we refer to an automated system that matches drivers and travelers on very short notice or even en route [1]. Current ridesharing platforms are Internet-based agencies (e.g., Carma, BlaBlaCar, CarpoolWorld, etc.) that match drivers' availabilities with passengers' needs in a centralized way [13]; in this scenario, the main challenges are to find optimal matches and immediately satisfy on-demand requests to form ridesharing instantaneously [1].

Recent studies [18], however, report a significant and continuous decline of ridesharing (i.e., ridesharing accounted for 19.7% of work trips in 1980 whereas it fell down to 10% in 2009). Why the immediacy and ubiquity of Internet-enabled mobile communications have not shifted people's choices of transportation has been recently studied in [1, 13]. Furuhata et al. [13] identify several difficulties that prevent ridesharing from being widely adopted. Building trust among travelers that do not know each other stands out as a crucial issue. It is well-known that most people are reluctant to travel with complete strangers (e.g., according to the survey in [7], only 7% of respondents would accept rides from strangers, whereas 98% and 69% would accept rides from a friend and the friend of a friend, respectively). The user's experience is usually the most reliable foundation of trust, and it can be spread through the network with a reputation system [10] by which users (both drivers and passengers) provide feedback on each other. Even though some matching agencies (such as Carma or Carpool World) already implement reputation systems, there is another key issue that severely hampers the ridesharing model offered by these agencies: users' privacy [13]. Indeed, a significant loss of privacy occurs as a result of matching agencies systematically collecting travel data and reputations of their customers [2], an information that may be sold or used for other purposes, such as marketing products or personal profiling [25].

To neutralize the above trust-related disincentives to ridesharing without running into privacy-related disincentives, we propose a fully decentralized P2P ridesharing system with the following features [21]:

- Being fully decentralized, it circumvents central matching agencies that may compile, aggregate and exploit individuals' data (location, travel habits, reputations) or that may bias ride matches because of commercial interests. Moreover, decentralization eliminates a central point of failure that might be the target of external attacks.
- For the sake of user privacy, we maintain information disclosure at a minimum, so that only the driver and the passenger(s) whose trips match learn each other's identity, desired trip and reputation.
- To tackle the reluctance of users to share trips with strangers, we implement a fully distributed reputation management mechanism whereby (i) drivers and passengers can rate each other after a shared ride, and (ii) peers can learn the reputation of others (resulting from aggregation of ratings of past rides) in a trustworthy way before deciding whether to share or not a future trip. Even though decentralized reputation may look more complex and less reliable than centralized reputation maintained by a trusted party, our reputation mechanism is designed in such a way that rational agents are interested in honestly cooperating to maintain it [10]. As a result, it is robust against a number of tampering attacks (e.g., fake reputation reporting, creation of multiple accounts, etc.), because agents cannot derive any benefit from such attacks.
- Our system rests on the theoretical foundations of co-utility [11, 12] to characterize users' collaboration and ensure that rational users (even purely selfish ones) will follow the proposed protocols.

The technical emphasis of our work is thus on the decentralized, privacy-preserving and efficient management and matching of drivers' offers and passengers' requests, as well as on the distributed calculation of peers' reputations. By designing the protocols in our system to be co-utile, we make sure the system will work as intended even if there is no central entity controlling its operation.

2 Decentralized P2P Ridesharing

Our system follows the *organized ridesharing* paradigm, by which ride matching can be prearranged between participants (unlike hailing a cab or hitchhiking, which are ad hoc). However, it does not necessarily assume previous involvements between the participants [8] (i.e., passengers and drivers do not need to know each other).

Organized ridesharing has been operated by agencies that match passengers and drivers according to a specific matching algorithm [13]. Ridesharing matches are two-sided, since they are performed according to the offers and requests received from drivers and passengers, respectively. Our system provides the same type of ridesharing matching service, but without relying on a central matching agency. Specifically, we offer a fully decentralized P2P network in which passengers and drivers act as peers that dynamically enter the network to offer or request rides. Ride matches are managed by the peers themselves in such a way that only those peers whose ride offers/requests match disclose their personal details (i.e., specific locations and identities) to each other. This feature overcomes the privacy concerns that have hampered ridesharing services thus far [13], because users are no longer forced to disclose all their ride details to a centralized agency.

2.1 Ridesharing Types

Ridesharing matching depends on the spatial and temporal features of ride offers and requests. In our system, each driver D has a trip route $R(D) = \{(l,t)_0^D, \ldots, (l,t)_n^D\}$ that is defined by an origin location and time $(l,t)_0^D$, a destination location and time $(l,t)_n^D$ and, optionally, an ordered set of the intermediate positions and associated (estimated) times $(l,t)_i^D$. Even if intermediate positions are optional, the more detailed the route (i.e., the more numerous and precise the intermediate points), the higher the chance to successfully match drivers' routes and passengers' requests, as it will be discussed below. Each passenger P, on the other hand, states his ride request by just specifying an origin location and desired pick-up time, and a destination, that is, $R(P) = \{(l,t)_0^P, (l)_1^P\}$. We assume passengers have no specific route constraints between their origins and destinations (the specific path followed to reach destination does not matter to them).

Similar to the usual practice by matching agencies, we do not require ride offers and requests to perfectly match (w.r.t. locations and times), because this would be too rigid and would severally restrict feasible matches. Instead, each passenger P defines spatial and temporal slacks that specify the maximum distance δ^P he accepts to walk (from his origin location l_0^P to the pick-up location by a driver, say l_i^D, and from the driver's drop-off location, say l_j^D, to his destination l_1^P), and the maximum time τ^P he accepts to wait before being picked up (with respect to his initial preference t_0^P). Our system supports the usual types of ridesharing implemented by state-of-the-art matching agencies [13, 15]:

- *Identical ridesharing*: the origins and destinations of driver and passenger match both spatially and temporally (within the passenger's space and time slacks); that is, $|l_0^D - l_0^P| \leq \delta^P \wedge |l_n^D - l_1^P| \leq \delta^P \wedge |t_0^D - t_0^P| \leq \tau^P$.
- *Inclusive ridesharing*: the origin and destination of the passenger are included in the route $R(D)$ of the driver (within the passenger's space and time slacks); that is, $\exists (l,t)_i^D \in R(D) \mid |l_i^D - l_0^P| \leq \delta^P \wedge |t_i^D - t_0^P| \leq \tau^P \wedge \exists l_j^D \in R(D) \mid$

$|l_j^D - l_1^P| \leq \delta^P$. With this ridesharing type, the driver does not need to deviate from her route, but she must stop at the pick-up and drop-off locations agreed with the passenger (l_i^D and l_j^D, respectively). As stated above, the more numerous and precise the intermediate positions/times of the driver's route, the higher the chance of finding an inclusive match with a passenger.

The above ridesharing types also support multiple passengers per vehicle (up to the maximum capacity c^D specified by each driver D for her vehicle). In inclusive ridesharing, the only difference between carrying one or several passengers is that in the latter case the driver must stop several times for pick-ups and drop-offs.

The literature also considers more complex ridesharing types. For example, in *partial ridesharing* the shared ride covers only a part of the passengers's trip, so that additional shared rides or transportation means need to be concatenated to complete the trip; in *detour ridesharing*, the driver may deviate from her route to accommodate the passengers' requests, at the expense of increased travel costs. Because these complex ridesharing types hinder instantaneous decision making by drivers and passengers and/or may require adding compensations for detouring, they are not currently offered by matching agencies [13] and we leave them as future work.

2.2 The P2P Ride Management Network

Rather than using a central matching agency, our system consists of a fully decentralized P2P network of drivers and passengers. Since decentralized P2P networks lack a trusted authority and a common "legal" framework, our challenge here is to design a co-utile protocol for ridesharing (which in turn will rely on a co-utile protocol for reputation management), that is, a ridesharing protocol that is followed by rational peers because of the benefits it brings to them.

There are several alternative communications technologies to implement decentralized P2P applications (e.g., CAN [9], Chord [23], Pastry [4], etc.). Among these, Pastry [4] is especially interesting because of its proximity-based routing. It consists of a self-organizing overlay network that provides a fault-tolerant and load-balanced distributed hash table (DHT) for large-scale P2P applications. Each node (i.e., a P2P user) is identified by a unique, randomly generated *nodeId* from a 128-bit identifier space, which acts as a pseudonym that hides the real identity (IP address) of the user. The DHT is internally used to map an existing *nodeId* to its associated IP address. In this manner, a message can be anonymously routed to any network user (via anycast) by forwarding it a bounded number of hops, concretely, $< \log_{2^b} N$, where N is the number of nodes and b typically has the value of 4. Moreover, each node requires locally storing only $O(\log_{2^b} N)$ entries of the hash table. Proximity-based message routing is performed by sending each message to the numerically closest *nodeId* that the sender has in its local hash table. If the receiver is not the message addressee, it forwards the message to its closest *nodeId*, and the process is repeated until the message reaches its intended addressee.

2.3 Distributed Ride-Matching Protocol

To match their rides, drivers and passengers need to advertise their offers and requests, thereby incurring some information disclosure. As discussed above, privacy concerns about that disclosure may deter users from participating in the ride-matching protocol [13]. Thus, to ensure that the protocol remains co-utile (and thus self-enforcing for rational agents), we need to keep the disclosure of the peers' spatio-temporal features as low as possible.

Within a decentralized network, a straightforward way to look for ride matches is to broadcast to all active peers a message with the passengers' requests and drivers' offers. However, this incurs a large information disclosure, and it may be even worse for privacy than using a centralized matching agency. Moreover, broadcast is very inefficient because it unnecessarily floods the network with messages that are useless to most peers. Instead, we propose a mechanism predicated on a topic-based decentralized subscription system implemented by the network by which (i) users only communicate with those peers with *similar* travel needs/offers and (ii) their concrete spatio-temporal data are only disclosed if a potential match is found. The idea is that drivers who offer rides *subscribe* to *topics* according to the routes they offer (topics are locally stored by the peers), and passengers *publish* their ride needs to the appropriate topic in the subscription system. Being decentralized, an event notification system for topic-based subscribe-publish applications is in charge of communicating via multicast only those peers whose rides match (i.e., drivers subscribed to a topic describing a route for which a passenger has published a message). Since the subscription information is local for each peer, topic subscriptions by drivers can only be known by publishing a message in a matching topic. Thanks to the proximity-based message routing and the topic-oriented multicast, the notification system is capable of scaling to a potentially large number of subscribers per topic and to a large number of different topics. Technically, the subscription infrastructure is built upon the decentralized Pastry P2P network to manage topic creation, topic subscription and dissemination of messages published in each topic.

Topics can be initially created (and subscribed to) by any peer (drivers, in our case) by defining a *topicId*. In our system, *topicId*s are defined according to the detail of the routes being offered by drivers, that is, all the possible rides consisting of a pick-up location/time and a drop-off location that can be defined from each route (that is, all subroutes of each route). Drivers offering rides equal to an existing *topicId* can register to the same topic to become subscribers of it. On the other hand, only peers knowing a *topicId* can publish messages in a certain topic. In our case, passengers publish messages in *topicIds* that match their ride needs (desired pick-up location/time and drop-off location). Thus, a passenger's message is disseminated only to the drivers subscribed to the matching topic, that is, to a ride offer that matches the passenger's needs; only in this case, peers may exchange their personal details (e.g., identities).

While the subscribe-publish mechanism routes the messages only to the peers whose offers/requests match and the anonymity of peers is preserved unless they

Fig. 1 Example of zone partitioning for the city of San Francisco: Zones correspond to neighborhoods

identify themselves to engage in a shared ride, topic subscription and publishing via detailed routes still discloses the spatio-temporal features of the users in the *topicIds*. Moreover, the number of different *topicIds* that can be defined according to *exact* locations and times may be unmanageably large. Also, too much detail may be counterproductive for the flexible matching described in Sect. 2.1: passengers ought to publish as many topics as locations and times included in their space and time slacks.

We tackle these issues by defining topics as *generalizations* of spatio-temporal features, rather than exact values. Specifically, exact locations l_i (both of passengers and riders) are generalized to zones within the area A (e.g. a city) where the ride-sharing system will be deployed. Zones are fixed by a predefined partition $Z(A)$ of the area (e.g., districts or neighborhoods in a city, uniform partitions in a map, etc.). Figure 1 shows an example of such partition for the city of San Francisco, where each $Z_i \in Z(A)$ corresponds to the name of a neighborhood. Likewise, concrete times t_i associated to positions l_i are generalized to fixed time intervals I (e.g., 10 min, a quarter of an hour, etc.). Specifically, since we measure t_i as the Epoch time (i.e., seconds since January 1st, 1970), the corresponding interval I is calculated as the integer part of the quotient between t_i and the size of the time interval (e.g., $\lfloor t_i/900 \rfloor$ for 15 min). Once generalized, each *topicId* is defined by the string resulting from

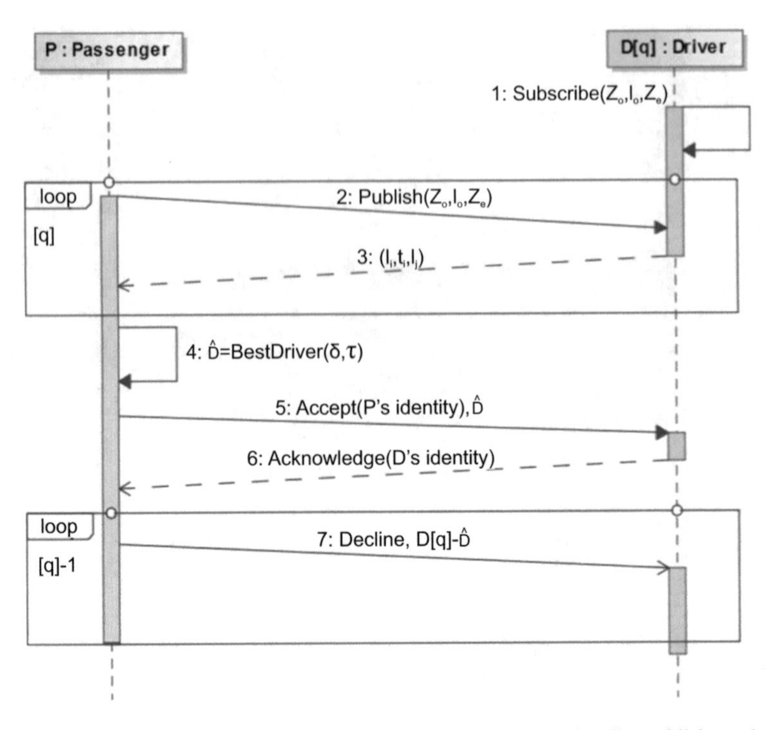

Fig. 2 Sequence diagram of the ride-matching protocol based on a subscribe-publish mechanism

concatenating the pick-up zone Z_o and time interval I_o with the drop-off zone Z_e. The system designer may configure beforehand the granularity of the generalization of both dimensions in order to control the level of incurred disclosure/matching accuracy. To optimize privacy, zones and time intervals should be taken to be as large as allowed by the spatio-temporal flexibility of passengers.

Formally, the ride-matching protocol (depicted in Fig. 2) is as follows:

1. Each driver D advertises the route $R(D)$ she would like to share by subscribing to as many topics as possible rides can be defined in her route. To do so, first, D generalizes positions and times $(l, t)_i^D \in R(D)$ to zones and time intervals $(Z, I)_i^D$, respectively; we refer to the *generalized route* as $\bar{R}(D) = \{(Z, I)_0^D, \ldots, (Z, I)_n^D\}$. Then, D subscribes to all topics $S(D)$ whose *topicIds* are triples (Z_o, I_o, Z_e), where $(Z_o, I_o) \in \bar{R}(D)$ is a generalized pick-up location-time pair and $Z_e \neq Z_o$ is a generalized drop-off location is such that $(Z_e, I_e) \in \bar{R}(D)$ for some generalized time interval I_e. Note that, by considering all the possible rides within the driver's route, we support the *inclusive ridesharing* described in Sect. 2.1.
2. Each passenger P looking for a ride (with desired route $R(P)$) publishes his request in a list of topics $S(P)$ that includes *topicIDs* obtained in a way analogous to the *topicIds* subscribed by a driver (see above). Specifically, since $R(P) = \{(l, t)_0^P, (l)_1^P\}$, then $S(P) \supseteq \{(Z_o, I_o, Z_e)\}$ where Z_o and Z_e are the zones

corresponding to the generalization of l_0^P and l_1^P, respectively, and I_o is the generalized time interval of t_0^P. If the passenger has flexible spatio-temporal requirements, $S(P)$ contains as many topics as zones and time intervals can be obtained by generalizing the locations and times resulting from adding/subtracting the passenger's space slack δ^P to locations in $R(P)$ and the passenger's time slack τ^P to the time in $R(P)$. As a result of the publication on these topics, only those drivers that have subscribed to the same topics will receive the passenger's request.

3. The drivers that receive a passenger's request (say q different drivers) respond with a direct message that discloses the precise pick-up and drop-off locations and pick-up time they generalized into the topic, that is, the specific $(l, t)_i^D$ and l_j^D.
4. Passenger P collects all the answers of the q matching drivers and selects the ride (and the respective driver) that specifically matches his request considering also his spatio-temporal flexibility δ^P and τ^P, as detailed in Sect. 2.1. If several rides (drivers) match his request, P selects the closest one (in normalized average space and time to his origin and destination) and sends the driver $\hat{D}$ offering that closest ride an "accept" message that reveals P's identity.
5. The driver $\hat{D}$ receiving the "accept" sends an "acknowledgment" to passenger P also revealing $\hat{D}$'s identity.
6. Finally, passenger P sends to the remaining $q - 1$ drivers collected at Step 4 a "decline" message. If the best-matching driver $\hat{D}$ does not reply to passenger P for some reason (e.g., it already received an "accept" from another passenger and reached the maximum occupancy c^D of her vehicle), P returns to Step 4 to select the next best alternative.

To illustrate the above-described protocol, assume a driver D wants to share the route $R(D)$ across the north of San Francisco shown in Fig. 3. $R(D)$ starts in Richmond, ends in Nob Hill, and consists of seven points listed in Table 1.

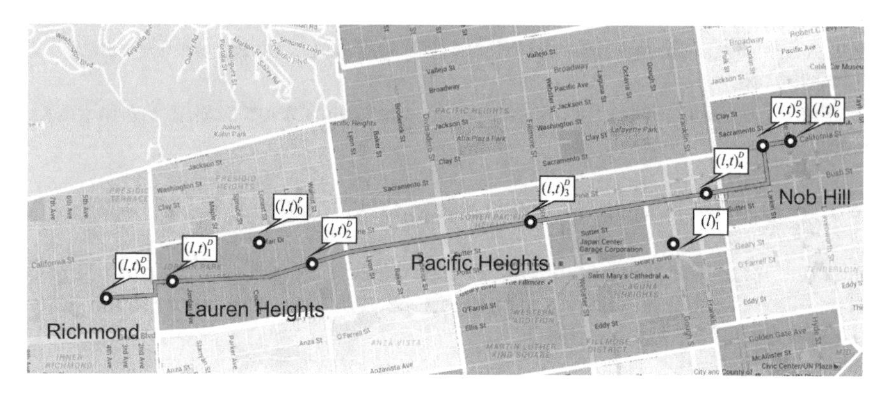

Fig. 3 Example driver D's route, $R(D)$, with 7 points, and (flexible) origin and destination for a passenger P

Table 1 Points (locations coordinates and Epoch times) of the example route $R(D)$

$R(D)$	Location coordinates (l)	Epoch time (t)
$(l, t)_0^D$	37.783094, −122.462326	1455534795
$(l, t)_1^D$	37.784009, −122.457863	1455534882
$(l, t)_2^D$	37.784891, −122.448550	1455535206
$(l, t)_3^D$	37.786892, −122.434302	1455535487
$(l, t)_4^D$	37.788419, −122.422887	1455535851
$(l, t)_5^D$	37.790759, −122.419025	1455536448
$(l, t)_6^D$	37.791166, −122.417394	1455536552

Table 2 Generalized route $\bar{R}(D)$ using San Francisco neighborhoods as zones and time intervals of 10 min

$\bar{R}(D)$	Location zones (Z)	Time intervals (I)
$(Z, I)_0^D$	Richmond	2425891
$(Z, I)_1^D$	Lauren heights	2425891
$(Z, I)_2^D$	Lauren heights	2425892
$(Z, I)_3^D$	Pacific heights	2425892
$(Z, I)_4^D$	Pacific heights	2425893
$(Z, I)_5^D$	Nob hill	2425894
$(Z, I)_6^D$	Nob hill	2425894

Route $R(D)$ is first generalized to $\bar{R}(D)$ (as shown in Table 2) according to the zone partition $Z(A)$ of San Francisco shown in Fig. 1 and time intervals of 10 min (i.e., $\lfloor t_i/600 \rfloor$).

To advertise the route, driver D subscribes to as many topics $S(D)$ as unique ordered combinations of pick-up/drop-off points can be derived from his generalized route $\bar{R}(D)$. Table 3 lists the triples (Z_o, I_o, Z_e) that constitute the *topicIds* D subscribes to.

Finally, let us consider that passenger P is looking for a ride with the following features

$$R(P) = \{(37.785146; -122.450288), 1455535117, (37.787215; -122.423938)\},$$

each one respectively corresponding to the points $(l, t)_0^P$ and $(l)_1^P$ shown in Fig. 3. Then, the passenger publishes a message in the *topicId* corresponding to the generalized points of his request, that is,

$$S(P) = (LaurenHeights, 2425891, PacificHeights).$$

Table 3 Set $S(D)$ of *topicIds* derived from the generalized route $\bar{R}(D)$ that driver D subscribes to

Z_o	I_o	Z_e
Richmond	2425891	Lauren heights
Richmond	2425891	Pacific heights
Richmond	2425891	Nob hill
Lauren heights	2425891	Lauren heights
Lauren heights	2425891	Pacific heights
Lauren heights	2425891	Nob hill
Lauren heights	2425892	Pacific heights
Lauren heights	2425892	Nob hill
Pacific heights	2425892	Pacific heights
Pacific heights	2425892	Nob hill
Pacific heights	2425893	Nob hill
Nob hill	2425894	Nob hill

There is a match between $S(D)$ and the 5th *topicId* subscribed to by D (Table 3). Therefore, the driver answers with the concrete details of that part of her ride, that is,

$$(l_2^D, t_2^D, l_4^D) = ((37.784891, -122448550), 1455535206, (37.788419, -122.422887)).$$

Since the driver's locations and time are within the passenger's spatio-temporal flexibility range given by δ^P and τ^P, the match is successful.

Note that all the steps of the protocol can be automatically managed without the intervention of the users. In Step 1, once drivers have defined their routes, subscription to the appropriate topics is deterministic and straightforward. Moreover, even if drivers just define the starting location and time and the destination of their rides, the system may also estimate the intermediate points and times of the route using a standard route planning algorithm. In Step 2, multicast messages are also automatically managed by the underlying network. In Step 3, response messages can be automatically created by de-generalizing zones and times and in Steps 4 to 6, the best match can be straightforwardly agreed upon according to the passenger restrictions.

A driver D can also unsubscribe from topics when, for some reason, her route changes or when the maximum capacity c^D of her vehicle is reached. Likewise, subscriptions are automatically removed from the system when the ride's time interval I is reached.

2.4 Co-utility Analysis of the Ride-Matching Protocol

When travel time and cost are the only utilities relevant to drivers and passengers, ridesharing always results in mutual benefits provided that travel costs (tolls, gas, etc.)

are split among the driver and the passenger(s) in a fair way, that is, so that the split cost is less than the cost of traveling alone for both the driver and the passenger(s).

Additionally, if high occupancy vehicles, carrying two or more passengers, are favored by the administration, then not only the cost may be further reduced for the driver and the passenger(s) (e.g. thanks to reduced highway tolls and/or parking spaces at reduced fees), but the travel time is likely to be shorter with respect to traveling alone (e.g. thanks to HOV lanes).

Since both drivers and passengers obtain benefits by sharing a ride (assuming their time/location/route constraints match), it is in the best interest of the agents to collaborate (i.e., to follow the ride-matching protocol). Hence, ridesharing is *co-utile*. In fact, because our matching patterns do not consider detours by drivers from their routes, co-utility in ridesharing only depends on the availability of agents with appropriate types, that is, compatible ride offers and requests. Therefore, provided that a large and balanced enough (with regard to ride offers/requests) community of drivers/passengers exists, ridesharing will be adopted by *rational* agents in a self-enforcing way.

As discussed in the introduction, (co-utile) ridesharing also boosts societal welfare in the medium term, because it reduces traffic congestion (thereby shortening travel time) and CO_2 emissions. Even though these societal benefits may further motivate some people to adopt ridesharing, the immediate individual cost and time benefits discussed above are arguably the most powerful incentives to collaboration.

However, at a closer look, travel time and cost are not the only utilities involved in ridesharing. If they were, ridesharing would be far more popular than it is nowadays. As discussed above, privacy concerns related to disclosing personal spatio-temporal information to network peers [7] or to matching agencies [2] may partly or totally offset the time/cost utilities, thereby thwarting co-utility and hence ridesharing. In our system, we minimize by design these "negative" privacy-related utilities because: (i) we do not rely on any central agency that systematically collects private data; (ii) users remain anonymous behind a randomly assigned *nodeId* and do not disclose their identities until a successful match is found; (iii) exact travel data are only exchanged between those peers whose ride offers/requests match (via subscription-driven multicast); and (iv) drivers' subscriptions in the P2P network only provide generalized information about their routes that is learned by publishing a request in a matching topic. One may certainly conceive systematic attacks whereby malicious users, behaving as passengers (resp. drivers), publish in (resp. subscribe to) topics with fake offers (resp. requests). However, since this attack should be systematic in order to obtain a significant amount of detailed data (i.e., complete drivers' routes, lists of passengers), peers can easily detect the attack pattern and decline communication with suspect peers.

As discussed in previous studies [10, 13], reluctance of drivers and passengers to share a trip with strangers is an even greater hurdle than privacy to the adoption of ridesharing. A natural way to mitigate users' mistrust when they lack direct experiences with their peers is to use a *reputation system*. Reputation, which captures the opinion of the community on each agent has at least two positive effects [10]:

- Reputation allows agents to build *trust*, which can neutralize the negative utilities related to mistrust (e.g., fear or reluctance in front of strangers). The higher an agent's reputation, the more trusted she is by other agents.
- Reputation makes agents accountable for their behavior: if an agent misbehaves (e.g., during the shared ride or by abusing/attacking the ride-matching system as described above), his reputation worsens and the other agents mistrust him more and more and are less and less interested to interact with him. In this manner, malicious agents (who may try to subvert the system, even irrationally) may be identified (via a low reputation) and penalized (e.g., through limitation or denial of service).

In the next section, we describe our proposal to incorporate a distributed reputation management protocol into our decentralized P2P network. The protocol we use is carefully designed so that it is itself co-utile, and hence even purely selfish agents are interested in following it. In this way, this protocol can be seamlessly used as a mechanism to enforce co-utility in protocols in which negative utilities would otherwise rule it out [10].

3 Distributed Reputation Management

In ridesharing, reputation is understood as the mutual feedback by drivers and passengers on each other. Such feedback may consist of the aggregation of several objective and subjective outcomes (e.g., how pleasant the trip was or whether there were any unnecessary delays or overcosts). This reputation model is inherited from commercial platforms (e.g. eBay), that successfully support buyers and sellers in building trust to unknown peers and eliciting honest behaviors [20].

Even though some ridesharing agencies (e.g., Carma, Carpool World, Golco) have incorporated reputation systems, they always manage reputations in a centralized way. Centralized reputation is straightforward to implement, but it adds to the privacy loss caused by centralized ridesharing: agencies not only learn where users go, but also what their reputations are.

To overcome this problem in a way that is coherent with the decentralized nature of our ridesharing management system, we propose a fully decentralized reputation management protocol. Moreover, our protocol should ensure that it is in the best interest of rational agents to follow it and report truthful reputation information (this is in fact the most critical issue in reputation systems [14]). In other words, *the reputation management protocol should be itself co-utile*. This aspect is crucial, if we want to use reputation management as a catalyst to turn ridesharing co-utile by neutralizing negative utilities (i.e., concerns about strangers).

The theoretical notions about reputation management and distributed reputation calculation can be found in [10] and also in the second chapter of this book.

3.1 Incorporating Reputations into the Ride-Matching Protocol

In ridesharing interactions, the lack of trust between driver and passenger is mutual and may hamper collaboration (i.e., drivers may be reluctant to share their cars with strangers, and passengers may be concerned with sharing a ride with a driver they do not know). With our reputation calculation mechanism, once global reputation values have been computed, it is immediate for any agent $\mathscr{P}_i$ (either driver or passenger) to learn the reputation value g_j of $\mathscr{P}_j$ (passenger or driver, respectively) in order to decide whether to ride with $\mathscr{P}_j$ or not. Agent $\mathscr{P}_i$ can query the M score managers of $\mathscr{P}_j$ for the latter's reputation; the resulting M values obtained from the M score managers should be the same, because the inputs of the score managers are the same. However, if some values differ (e.g., if some score managers or agents involved in the calculation have altered the computation for some -non-rational/random- reason), $\mathscr{P}_i$ can take as g_j the most common value among the ones sent by the score managers.

In addition to the global reputation of $\mathscr{P}_j$, agent $\mathscr{P}_i$ may also rely on his direct experiences with $\mathscr{P}_j$, if any, which are reflected in his local normalized value c_{ij}. In some cases, local and global reputation values may not be coherent because the latter are the aggregated version of the former. Thus, for a more robust trust enforcement, it is better for $\mathscr{P}_i$ to consider both local reputations and global reputations and take the lowest value in order to make decisions about collaboration. In this manner, agents will be discouraged from selectively behaving well with some agents while misbehaving with others.

Additionally, individual agents may have different levels of reluctance versus strangers and, thus, may require different reputation levels to build trust and agree to collaborate with other agents. To model this notion, we may allow each agent $\mathscr{P}_i$ to specify a *minimum reputation* threshold ρ_i that any $\mathscr{P}_j$ should have in order for $\mathscr{P}_i$ to ride with $\mathscr{P}_j$. Since ridesharing is predicated on reciprocal trust, it will only happen if both the driver's reputation g_D and the passenger's reputation g_P are above the other party's reputation threshold (i.e., $(g_D \geq \rho_P) \wedge (g_P \geq \rho_D)$).

In summary, the ride-matching protocol detailed in Sect. 2.3, can be easily extended to enforce trust among the agents by considering the peers' reputations in Steps 3 and 4. In Step 3, the q drivers receiving a passenger P's request may evaluate P's reputation and, only if it is above their desired respective reputation thresholds, the drivers will answer with their respective ride details. Likewise, in Step 4, passenger P will select the best ride (driver $\hat{D}$) only from those drivers with a reputation above his desired reputation threshold ρ^P. In terms of information disclosure, we can also see that with this reputation-aware protocol a passenger and a driver reveal their identities to each other (in Steps 4 and 5, respectively) *only if they both trust each other*. Finally, as discussed in Sect. 2.4, peers' reputations can also be used to punish agents that are suspected of abusing/attacking the system. All these enhancements brought by the reputation mechanism make following the ride-matching protocol the only rational choice and, thus, they enforce co-utility even if peers do not know each other.

4 Empirical Study

This section reports the results of an empirical study that simulates realistic rides in our P2P network. Data about rides have been extracted from the cab mobility traces provided by the Exploratorium museum within the cabspotting project.[1] This data set contains the traces of the trips of approximately 500 cabs during May 2008 in San Francisco Bay Area [19]. Each trace is defined as the GPS coordinates and the absolute times of its origin, destination and intermediate positions measured every 10 seconds (on average).

4.1 Configuration of the Experiments

In our experiments we used the mobility traces of a whole week to simulate ride offers (by drivers) and requests (by passengers). We considered only those traces in which the cab was occupied by a customer, because these are the ones corresponding to realistic routes with meaningful and precise origins and destinations. In contrast, we omitted cabs wandering in search of customers, because this may result in seemingly random routes. We also omitted very short routes of less than 500 m (for which sharing a ride is not worth while) and/or those with less than 4 position measurements. The resulting set contains 94,070 traces. Figure 4 (left) shows that the traces are uniformly distributed through the week (with a small dominance of Saturdays), whereas Fig. 4 (middle) shows that they are concentrated during working hours ([8−18 h]) and the leisure time ([19−1 h]). Likewise, Fig. 4 (right) shows that the length of most rides is 3 km or less.

We simulated three well-differentiated scenarios by randomly assigning a percentage of the traces to drivers (ride offers) and passengers (ride requests), as follows:

- *Balanced scenario*: 50% of the traces (47,035) are assigned to drivers' offers and 50% (47,035) to passengers' requests.
- *Driver-dominated scenario*: 70% (65,849) of the traces are assigned to drivers' offers and 30% (28,221) to passengers' requests. This is a favorable setting for passengers, since they can choose from a wide range of offers.
- *Passenger-dominated scenario*: 30% (28,221) of the traces correspond to drivers' offers, whereas 70% (65,849) correspond to passengers' requests. This favors drivers, since they can choose from more passengers. Since there are less drivers than passengers, ridesharing may involve sharing the same trip with several passengers, which also contributes to reducing travel expenses for each ridesharer.

For each driver D, we used all the measurements of her mobility trace (positions and times) in the data set to define her route $R(D)$. For each passenger P, we used only the initial position and time of his trace to define l_0^P and t_0^P, respectively, and the final position to define l_1^P. We also set different levels of passenger flexibility regarding

[1] http://www.exploratorium.edu/id/cab.html.

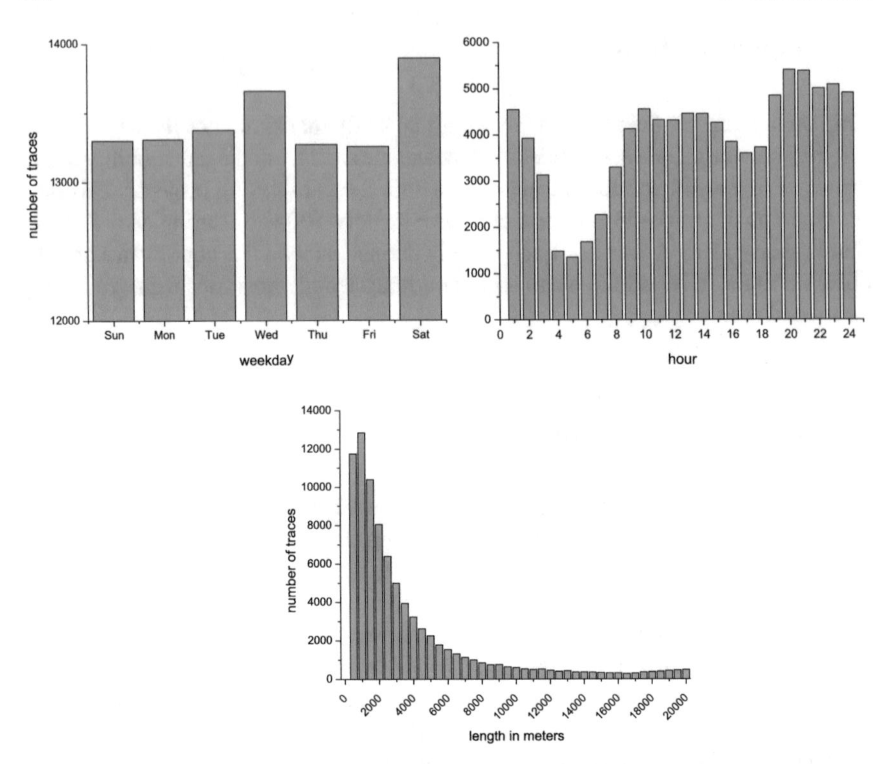

Fig. 4 Distribution of the mobility traces used in our experiments: by days of the week (*top left*), by hours of the day (*top right*) and by length (*bottom*)

the pick-up/drop-off locations and times: the maximum distance δ^P that a passenger P will be willing to walk (with respect to the pick-up and drop-off locations) has been set to $\delta^P = \{250, 500, 750, 1000\}$ m, whereas the maximum waiting time τ^P for pick-up has been set to $\tau^P = \{5, 10, 15, 20\}$ min. The maximum values of both dimensions are coherent with recent surveys on carpooling [3], which show that ride partners are usually found within 1 km radius (roughly 20 min walk time) of their current locations.

The management and matching of rides have been implemented as detailed in Sect. 2.2: drivers define their ride offers in a privacy-preserving way by subscribing to the set of topics defined by the city zones and generalized time intervals corresponding to their routes, and receive requests from potential passengers whose (generalized) ride requirements match. City zones for San Francisco Bay have been taken to be neighborhoods and generalized time intervals have been taken as 10-min intervals. Because the sub-routes defined by the intermediate positions of the rides are also considered (during the subscription and ride matching), both the *identical* and the *inclusive* ride matching patterns detailed in Sect. 2.1 have been implemented. Finally, we set a maximum occupancy per vehicle of $c^D = 4$ passengers (plus the driver).

The results of the simulations for the different scenarios have been evaluated according to the following metrics:

- Percentage of passengers that found a matching driver's offer. This metric reflects the functional success of ridesharing for passengers because, if passengers cannot find drivers, they will be forced to look for an alternative conveyance (e.g., public transportation, cab or their own private vehicle).
- Average occupancy (in number of passengers other than the driver) of the vehicles. This represents the economic success of ridesharing, both for the driver (the higher the occupancy, the more people she can share costs with) and also for the passengers (who will each bear a lower cost). For simplicity, we measured the occupancy as the number of different passengers that share (a part of) the ride through the entire driver's route.
- Percentage of kms saved by sharing rides compared to all agents (both passengers and drivers) traveling separately (for example, with their own private vehicles). This metric measures the boost of social welfare resulting from ridesharing, since saved travel kms imply a reduction of CO_2 emissions.

4.2 Results

Figure 5 shows the results for the three scenarios introduced above according to the three evaluation metrics. Regarding the percentage of passengers that found a matching ride, results are proportional to (i) the ratio of passengers/drivers (i.e., the more drivers are available, the higher the chance of finding a match) and (ii) the passengers' flexibility regarding time and locations. Flexibility turns out to have the greatest influence on the results, as the percentage of matches increases by 6–9 times between the least flexible and the most flexible requirements (i.e., 15 vs. 87%, 12 vs. 78% and 8 vs. 70% for the driver-dominated, balanced and passenger-dominated scenarios, respectively). Since most ridesharing solutions available in the literature [13] assume that passengers and drivers are matched by proximity rather than by exact position/times, we can see that (flexible) ridesharing has a high potential.

Regarding vehicle occupancy, results are opposite: the more passengers available, the more occupied are vehicles, which results in more saving for everyone. Differences are more significant when comparing the different scenarios: an average 1.623 passengers/vehicle is achieved in the passenger-dominated scenario versus an average 0.355 in the driver-dominated one. However, even the higher average occupancy is still far from the maximum occupancy of 4 passengers/vehicle. Like matchings, occupancy increases very much with the passengers' flexibility, specifically by a factor 6–9 between the least flexible and the most flexible requirements.

The percentage of kms saved as a result of sharing rides is perfectly proportional to the number of passenger matches (the more matches, the more passengers avoid resorting to their own private cars). For the driver-dominated scenario with

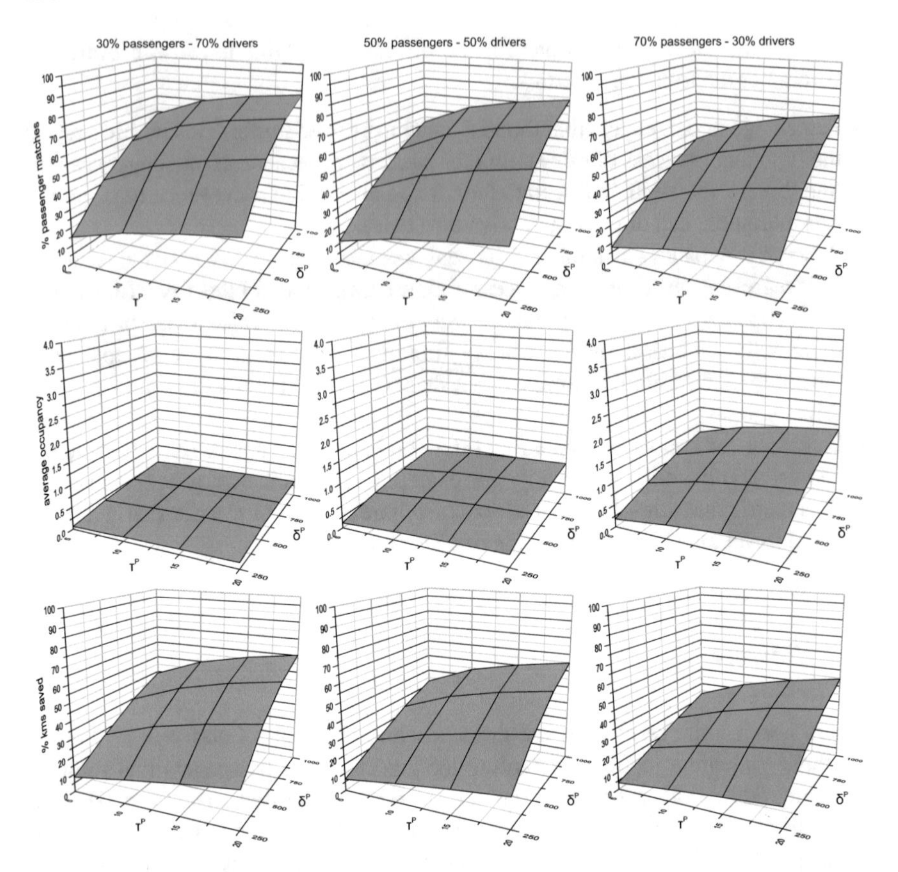

Fig. 5 Percentage of passengers that found a matching ride (*top*), average occupancy of vehicles (*middle*) and percentage of kms saved (*bottom*) as a result of ridesharing for the three scenarios (*left* driver-dominated; *center* balanced; *right* passenger-dominated)

the maximum passenger flexibility, as much as 65% kms are saved (130,588 kms in absolute terms). In terms of social welfare and extrapolating the weekly data we considered in our experiments to a year period, saving 65% kms in this scenario represents an annual saving of around 739 thousand gas liters (assuming a consumption of 0.1089 l/km [16]) and a reduction of 1.73 tons of CO_2 emissions (assuming that 255.38 g of CO_2 are emitted per kilometer [17]).

4.3 Lack of Trust and Reputation Management

In the previous simulations, we have assumed that peers do not mind sharing rides with strangers. In this section, we drop this assumption and evaluate the effectiveness of the reputation management mechanism proposed in Sect. 3 to cope with mistrustful peers.

As a baseline, we focus on the most favorable scenario for passengers, the driver-dominated scenario with maximum passenger flexibility. In the second row of Table 4 we report both the percentage of passengers that found a matching ride (very high: 82.74%) and the average vehicle occupancy in the ideal *Trusted scenario* in which all agents trust each other (or at least do not mind sharing rides with strangers). However, as discussed in [7] only 7% of the respondents seem to be willing to accept rides from strangers. The third row of Table 4 simulates a completely *Untrusted scenario* in which only that 7% of passengers will accept rides with drivers and in which the effectiveness of ride-matching is severely hampered (only 5.79% of matches). Finally, the remaining rows simulate *Untrusted scenarios* in which agents rely only on reputations to decide whether to share or not a matching ride. In all cases, each agent $\mathscr{P}_i$'s global reputation g_i has been randomly chosen in the range [0, 1], which corresponds to the normalized reputation range of the calculation mechanism depicted in Sect. 3. Then, different scenarios have been defined by setting a minimum and equal reputation requirement ρ for all agents in the system, $\rho = \{0.1, 0.2, 0.4, 0.6, 0.8\}$; that is, if we set $\rho \geq 0.4$, by following the reputation-based protocol depicted in Sect. 3.1, a matching ride would only be accepted if both the driver and the passenger have a global reputation value $g \geq 0.4$. Because of the randomly assigned reputations, for consistency, all the reported results are the average of 5 runs. We can see that in this *Untrusted scenario*, the use of reputations builds trust among the agents and increases the effectiveness of ridesharing in a way that is inversely proportional to the reputation requirements of the agents.

Finally, we simulated a more realistic and heterogeneous scenario in which each agent $\mathscr{P}_i$ is assigned a random global reputation g_i and also a random reputation requirement ρ_i with respect to the other peers, both uniformly distributed within the reputation value range [0..1]. Then, we measured how the global reputation values of peers affected the success of passengers in finding a ride and the success of drivers in finding passengers. Moreover, for passengers we also measured the average distance they needed to walk from their origin to their pick-up location (and also from the

Table 4 Percentage of passengers that found and accepted a matching ride, and vehicle occupancy for different scenarios of trust and reputation requirements

Scenario	*P*'s matches (%)	*D*'s vehicle occupancy
Trusted	82.74	0.355
Untrusted (no reputations)	5.79	0.025
Untrusted ($\rho \geq 0.8$)	11.88	0.051
Untrusted ($\rho \geq 0.6$)	28.28	0.121
Untrusted ($\rho \geq 0.4$)	46.14	0.198
Untrusted ($\rho \geq 0.2$)	64.54	0.277
Untrusted ($\rho \geq 0.1$)	73.25	0.314

Table 5 Passengers' average waiting time and walked distance, and average driver's vehicle occupancy for matching peers according to their global reputation values

Peers' global reputation	P's wait. time (s)	P's walk. distance (m)	P's matches (%)	Vehicle occup. for matched D	D's matches (%)
[0.0...0.25)	478.99	297.95	13.83	1.09	5.97
[0.25...0.50)	409.25	293.60	24.84	1.24	17.61
[0.50...0.75)	368.17	280.20	29.15	1.41	30.30
[0.75...1.0]	342.36	268.70	32.18	1.65	46.12

drop-off location to their destination) and the time they had to wait for pick-up. For drivers we also measured the average vehicle occupancy for those drivers that were able to find a match. For both, this was measured according to their global reputation values. Results are shown in Table 5, both for passenger and drivers, for (global) reputation quartiles; again, they are the average of 5 simulation runs.

It is clear from Table 5 that, in an scenario with uniformly distributed reputations and reputation requirements, the higher the reputation of the peers, the higher their chance of finding other peers with matching rides that *trust* them. This is also reflected in the higher vehicle occupancy achieved by drivers with high reputations: they are able to find even more than one matching passenger. On the other hand, passengers with high reputations also fare better: they need to walk shorter distances to/from the pick-up/drop-off location and they wait shorter times for pick-up. The explanation is that passengers with higher reputations can select the best matching offer in spatio-temporal terms from a wider range of offers (at Step 4 of the ride-matching protocol detailed in Sect. 2.3).

This last aspect strengthens reputation as the mechanism to enforce co-utility in ridesharing: if both passengers and drivers are aware that a higher reputation will allow them to find more (and better) matches, it will be in their own interest to behave well with their peers in order to increase their reputations. In other words, *reputation becomes a new, secondary utility that agents wish to optimize*, along with the primary time and cost utilities. In turn, by promoting the good ridesharing behaviors that peers rate positively (e.g., avoid unnecessary delays or costs, or provide a pleasant company), we make peers' collaboration more harmonious and sustainable.

5 Conclusions

Coming up with mechanisms that contribute to the adoption of ridesharing is of great interest, both for end users and for the society at large. Indeed, ridesharing has the potential to bring significant benefits to the involved agents and a more sustainable management of transportation to society.

In this chapter we have tackled two of the main obstacles deterring ridesharing (i.e., lack of trust and privacy concerns with respect to matching agencies) by means of a reputation-enabled privacy-preserving decentralized P2P ridesharing network. Even though decentralization entails additional challenges in comparison with the usual solutions based on central matching agencies, our ride-matching and reputation management protocols have been carefully designed so that peers are rationally motivated to adhere to them. In addition to being self-enforcing, our protocols are also *co-utile*, because they bring a number of mutual benefits to the involved peers (e.g., reduced travel costs, fair reputation calculation, better ride matches, etc.).

The reputation management protocol is a clear example of how intrinsically co-utile incentive protocols can spark co-utility in other protocols that are not intrinsically co-utile [21, 24]: the reputation mechanism is self-enforcing and it enables co-utility in the ride-matching protocol, that would not be co-utile without reputation, due to the lack of trust. Moreover, as shown by our simulations, reputation can also enhance the benefits of collaboration (i.e., a higher reputation brings more and better matches). This makes reputation an additional utility that agents wish to maximize, which motivates them to behave well and results in smoother collaboration and more social welfare.

Acknowledgements Funding by the Templeton World Charity Foundation (grant TWCF0095/AB60 "CO-UTILITY") is gratefully acknowledged. Also, partial support to this work has been received from the Government of Catalonia (ICREA Acadèmia Prize to J. Domingo-Ferrer and grant 2014 SGR 537), the Spanish Government (projects TIN2014-57364-C2-1-R "SmartGlacis", TIN2015-70054-REDC and TIN2016-80250-R "Sec-MCloud") and the European Commission (projects H2020-644024 "CLARUS" and H2020-700540 "CANVAS"). The authors are with the UNESCO Chair in Data Privacy, but the views in this work are the authors' own and are not necessarily shared by UNESCO or any of the funding bodies.

References

1. Agatz, N.A.H., Erera, A.L., Savelsbergh, M.W.P., Wang, X.: Optimization for dynamic ridesharing: a review. Eur. J. Oper. Res. **223**(2), 295–303 (2012)
2. Amey, A.M.: A proposed methodology for estimating ridesharing viability within an organization, application to the MIT community. In: Transportation Research Board Annual Meeting (2011)
3. Buliung, R., Soltys, K., Bui, R., Habel, C., Lanyon, R.: Catching a ride on the inforamtion superhighway: toward an understanding of internet-based carpool formation and use. Transportation **37**(6), 849–873 (2010)
4. Castro, M., Druschel, P., Hu, Y.C., Rowstron, A. Exploiting network proximity in peer-to-peer overlay networks. Technical report MSR-TR-2002-82 (2002)
5. Caulfield, B.: Estimating the environmental benefits of ridesharing: A case study of Dublin. Transp. Res. D **14**, 527–531 (2009)
6. Chan, N.D., Shaheen, S.A.: Ridesharing in North America: past, present and future. Transp. Rev. **32**(1), 93–112 (2012)
7. Chaube, V., Kavanaugh, A.L., Pérez-Quiones, M.A.: Leveraging social networks to embed trust in rideshare programs. In: Proceedings of the Hawaii International Conference on System Sciences (HICSS), pp. 1–8 (2010)
8. Dailey, D.J., Loseff, D., Meyers, D.: Seattle smart traveler: dynamic ridematching on the world wide web. Transp. Res. C **7**(1), 17–32 (1999)

9. Deering, S.E., Cheriton, D.R.: Multicast routing in datagram internetworks and extended LANs. ACM Trans. Comput. Syst. **8**(2), 85–110 (1990)
10. Domingo-Ferrer, J., Farràs, O., Martínez, S., Sánchez, D., Soria-Comas, J.: Self-enforcing protocols via co-utile reputation management. Inf. Sci. **367**(C), 159–175 (2016)
11. Domingo-Ferrer, J., Martínez, S., Sánchez, D., Soria-Comas, J.: Co-utility: self-enforcing protocols for the mutual benefit of participants. Eng. Appl. Artif. Intell. **59**, 148–158 (2017)
12. Domingo-Ferrer, J., Sánchez, D., Soria-Comas, J.: Co-utility: self-enforcing collaborative protocols with mutual help. Prog. Artif. Intell. **5**(2), 105–110 (2016)
13. Furuhata, F., Dessouky, M., Ordóñez, F., Brunet, M.E., Wang, X., Koening, S.: Ridesharing: the state-of-the-art and future directions. Transp. Res. B **57**, 28–46 (2013)
14. Jurca, R., Faltings, B.: An incentive compatible reputation mechanism. In: Proceedings of the IEEE International Conference on E-Commerce (CEC), pp. 285–292 (2003)
15. Morency, C.: The ambivalence of ridesharing. Transportation **34**(2), 239–253 (2007)
16. Office of Highway Policy Information (2013) Annual Vehicle Distance Traveled in Miles and Related Data. https://www.fhwa.dot.gov/policyinformation/statistics/2013/vm1.cfm. Accessed 26 Feb 2016
17. Office of Transportation and Air Quality (2016) Measuring Greenhouse Gas Emissions from Transportation. http://www3.epa.gov/otaq/climate/measuring.htm. Accessed 26 Feb 2016
18. Park, H., Gebeloff, R.: Car-pooling declines as driving becomes cheaper. The New York Times. Accessed 29 Jan 2011
19. Piorkowski, M., Sarafijanovoc-Djukic, N., Grossglauser, M.: A parsimonious model of mobile partitioned networks with clustering. In: Proceedings of the International Conference on COMmunication Systems and NETworkS (COMSNETS) (2009)
20. Resnick, P., Zeckhauser, R.: Trust among strangers in internet transactions: empirical analysis of eBay's reputation system. Econ. Internet E-Commerce **11**, 127–157 (2002)
21. Sánchez, D., Martínez, S., Domingo-Ferrer, J.: Co-utile P2P ridesharing via decentralization and reputation management. Transp. Res. C **73**, 147–166 (2016)
22. Schrank, D., Lomax, T.: Urban mobility report. Texas Transportation Technical Report (2007)
23. Stoica, I., Morris, R., Karger, D., Frans Kaashoek, M., Balakrishnan, H.: Chord: A scalable peer-to-peer lookup service for internet applications. In: Proceedings of the 2001 Conference on Applications, Technologies, Architectures, and Protocols for Computer Communications (SIGCOMM '01), pp. 149–160. ACM (2001)
24. Turi, A.N., Domingo-Ferrer, J., Sánchez, D.: Filtering P2P loans on co-utile reputation. In: Proceedings of the 13th International Conference on Applied Computing, AC 2016, pp. 139–146 (2016)
25. U.S. Federal Trade Commission. Data Brokers, A Call for Transparency and Accountability (2014)

Self-enforcing Collaborative Anonymization via Co-utility

Jordi Soria-Comas and Josep Domingo-Ferrer

Abstract In surveys collecting individual data (microdata), each respondent is usually required to report values for a set of attributes. If some of these attributes contain sensitive information, the respondent must trust the collector not to make any inappropriate use of the data and, in case any data are to be publicly released, to properly anonymize them to avoid disclosing sensitive information. If the respondent does not trust the data collector, she may report inaccurately or report nothing at all. The reduce the need for trust, local anonymization is an alternative whereby each respondent anonymizes her data prior to sending them to the data collector. However, local anonymization by each respondent without seeing other respondents' data makes it hard to find a good trade-off minimizing information loss and disclosure risk. In this chapter, we detail a distributed anonymization approach where users collaborate to attain an appropriate level of disclosure protection (and, thus, of information loss). Under our scheme, the final anonymized data are only as accurate as the information released by each respondent; hence, no trust needs to be assumed towards the data collector or any other respondent. Further, if respondents are interested in forming an accurate data set, the proposed collaborative anonymization protocols are self-enforcing and co-utile [3, 5].

1 Introduction

A microdata file contains data collected from individual respondents. Due to their level of detail, microdata can be useful for a variety of secondary analyses by third parties other than the data collector. However, releasing the original data is not feasible because it would lead to a violation of the privacy of respondents. Statistical disclosure control (SDC), a.k.a. statistical disclosure limitation, for microdata seeks

J. Soria-Comas (✉) · J. Domingo-Ferrer
UNESCO Chair in Data Privacy, Department of Computer Science and Mathematics,
Universitat Rovira i Virgili, Av. Països Catalans 26, 43007 Tarragona, Catalonia, Spain
e-mail: jordi.soria@urv.cat

J. Domingo-Ferrer
e-mail: josep.domingo@urv.cat

J. Domingo-Ferrer and D. Sánchez (eds.), *Co-utility*, Studies in Systems,
Decision and Control 110, DOI 10.1007/978-3-319-60234-9_8

to produce an anonymized version of the microdata file such that it enables valid statistical analyses but thwarts inference of confidential information about any specific individual [18].

The mainstream literature on SDC for microdata (e.g. see [10]) focuses on centralized anonymization, which features a trusted data collector. The data collector (e.g. National Statistical Institute) gathers original data from the respondents and takes care of anonymizing them. While avoiding the computational burden of anonymization is confortable to respondents, it has the downside that they need to trust the data collector. Some data collectors may be trusted (e.g. national statistical agencies, which must protect the privacy of individuals under a legal mandate) but, as data collection has become pervasive, it is unreasonable to believe that all collectors can be trusted [7, 21].

Local anonymization is an alternative disclosure limitation paradigm suitable for scenarios where the respondents do not trust (or trust only partially) the data collector. Each respondent anonymizes her own data before handing them to the data collector. In comparison to centralized anonymization, local anonymization usually results in greater information loss. The reason is that each respondent needs to protect her data without seeing the other respondents' data, which makes it difficult for her to find a good trade-off between the disclosure risk limitation achieved and the information loss incurred.

To overcome the limitations of the centralized and the local anonymization paradigms, we propose the notion of *collaborative anonymization* [22], which is in line with the novel notion of co-utility [3, 5] (see also the first chapter of this book). The advantage of co-utility is that it leads to a system that works smoothly without the need of external enforcement.

2 Related Work

This work seeks to empower each respondent to anonymize her own data while preserving utility as in the centralized paradigm.

Related works exist that consider privacy-conscious data set owners, rather than privacy-conscious respondents. When dealing with privacy-conscious data set owners, one faces a data integration problem where the data owners do not want to share data that are more specific than those in the final anonymized data set to be jointly obtained. In [23] a top-down generalization approach for two owners of vertically partitioned data sets is proposed. Both owners start with the maximum level of generalization, and they iteratively and collaboratively refine the generalization. In [11, 12] the same problem is tackled by using cryptographic techniques. In [13] the anonymization of horizontally partitioned data sets is considered. The main difference between the above proposals and our work is that the number of respondents is usually much greater than the number of data set owners (the latter are a small number in most realistic data integration settings). In our case, there is a different respondent for each data record being collected, which makes proposals oriented to a few data set owners unusable.

Among the related works specifically addressing respondent privacy, the local anonymization paradigm is closest to our approach in terms of trust requirements. Several local anonymization methods have been proposed. Many basic SDC techniques such as global recoding, top and bottom coding, and noise addition can be applied locally (see [6, 10] for details on such techniques). There are, however, some techniques specifically designed for local anonymization that, in addition to helping a respondent hide her response, allow the data collector to get an accurate estimation of the distribution of responses for groups of respondents. In randomized response [24], the respondent flips a coin before answering a sensitive dichotomous question (like "Have you taken drugs this month?"); if the coin comes up tails, the responder answers "yes", otherwise she answers truthfully. This protects the privacy of respondents, because the survey collector cannot determine whether a particular respondent's "yes" is random or truthful; but he knows that the "no" answers are truthful, so that he can estimate the real proportion of "no" as twice as much as the observed proportion of "no" (from which the real proportion of "yes" follows). FRAPP [1] can be seen as a generalization of random response. In FRAPP, the respondent reports the real value with some probability and, otherwise, it returns a random value from a known distribution. In AROMA [19] each respondent hides her confidential data within a set of possible confidential values drawn from some known distribution. In any case, to obtain an accurate result, the output of a query performed on the anonymized data must be adjusted according to the known distribution used to mask the actual data. While some kind of adjustment of the query results may also be needed in the centralized paradigm (e.g. when the generalization used for quasi-identifiers in a k-anonymous data set does not match the query; the randomness introduced by local anonymization makes the estimate less accurate than in centralized anonymization.

An advantage of local anonymization, though, is that the respondent is given some capability to decide the amount of anonymization required, which is likely to increase her disposition to provide truthful data (rather than fake data). Yet, most privacy models/techniques give uniform disclosure limitation guarantees to all respondents, which may not suit the different perceptions of disclosure risk of the various respondents. To address this concern, [26] proposed a privacy model in which each individual determines the amount of protection required for her data.

3 Collaborative Anonymization: Requirements and Justification

A problem with centralized anonymization is that, if a respondent does not trust the data collector to properly use and/or anonymize her data, she may decide to provide false data (hence causing a response bias) or not data at all (hence causing a non-response bias). Local (also known as independent) anonymization is an alternative that is not free from problems either. Permutation is essential to anonymization, but the permutation caused by a certain amount of masking depends not only on one's own record but on the values of the records of the other respondents. Hence, for a respondent anonymizing her own record in isolation, it is hard to determine the amount of masking that yields a good trade-off between disclosure risk and information loss, i.e. that causes enough permutation but not more than enough permutation. A natural tendency is for each respondent to play it safe and overdo the masking, just in case, which incurs more information loss than necessary.

To deal with the above shortcomings of centralized and local anonymization, we propose a new paradigm that we call collaborative data anonymization. Consider a set of respondents $R_1, \ldots, R_m$ whose data are to be collected. Each respondent is asked to report information about a set of attributes (some of them containing confidential/sensitive information). Since respondents place limited trust on the data collector, they may refuse to provide the collector with non-anonymized data. A more realistic goal is to generate, in a collaborative and distributed manner, an anonymized data set that satisfies the following two requirements: (i) it incurs no more information loss than the data set that would be obtained with the centralized paradigm for the same privacy level, and (ii) neither the respondents nor the data collector gain more knowledge about the confidential/sensitive attributes of a specific respondent than the knowledge contained in the final anonymized data set.

In general, the motivations for a respondent to contribute her data are not completely clear. A rational respondent will only contribute if the benefit she gets from participating compensates her privacy loss. It is not in our hands to determine what the motivations of the respondents are. However, since our collaborative approach achieves the same data utility as the centralized approach while improving the respondent's privacy vs the data collector, any respondent willing to participate under the centralized approach should be even more willing to participate under our collaborative scheme. More precisely, we can distinguish several types of respondents depending on their interests in the collected data and in their own privacy:

- A respondent without any interest in the collected data set is better off by declining to contribute.
- A respondent who is interested in the collected data and has no privacy concerns can directly supply her data and needs no anonymization (neither local, nor centralized nor collaborative).
- A respondent who is interested in the collected data but has privacy concerns will prefer the collaborative approach to the centralized and the local approaches. Indeed, the collaborative approach outperforms the centralized approach in that the former offers privacy vs the data collector. Also, the collaborative approach outperforms the local approach in that it yields a collected anonymized data set with less information loss, that is, with higher utility.

Remark 1 (co-utile anonymization). Note that the level of privacy protection obtained by a respondent affects the privacy protection that other respondents get. A basic approach for preserving the privacy of a specific respondent is based on hiding that respondent within a group of respondents. None of the respondents in such a group is interested making any of the respondents in the group re-identifiable, that makes her own data more easily re-identifiable. For example, if one record in a k-anonymous group is re-identified, the probability of successful re-identification for the other group members increases from $1/k$ to $1/(k-1)$. This fact suggests that a respondent is interested not only in protecting her privacy, but also in helping other respondents in preserving theirs. This is the fundamental principle behind the notion of co-utility (see [3, 5] and the first chapter of this book): the best strategy to attain one's goal is to help others in attaining theirs. The fact that privacy protection turns out to be co-utile ensures that respondents will be willing to collaborate with each other to improve the protection of all the group.

4 Collaborative k-Anonymity

This section describes how to generate a k-anonymous data set in a distributed manner, such that none of the respondents releases more information than the one available on her in the final k-anonymous data set. To this end, some communication between the respondents is needed to determine the k-anonymous groups.

In general, there can be several combinations of attributes in a data set that together act like a quasi-identifier, that is, such that each combination of attributes can be used to re-identify respondents; for example, one might have a quasi-identifier *(Age, Gender, Birthplace)* and another quasi-identifer *(Instruction_level, City_of_residence, Nationality)*. Without loss of generality and for the sake of simplicity, we will assume there is a single quasi-identifier that contains all the attributes that can potentially be used in record re-identification. Note that this is the worst-case scenario. Let QI be the set of attributes in this quasi-identifier.

Quasi-identifier attributes are usually assumed to contain no confidential information, that is, the set of quasi-identifier attributes is assumed to be disjoint from the set of confidential/sensitive attributes. This assumption is reasonable, because it is equivalent to saying that the attacker's background information does not include sensitive information on any respondent (indeed, the attacker wants *to learn* sensitive information, so it is reasonable to assume that he does not yet know it). Certainly, there might be special cases in which the attacker knows and uses sensitive data for re-identification, but we will stick to the usual setting in which this does not happen.

Since the attributes in QI are non-confidential, respondents can share their values among themselves and with the data collector, so that all of them get the complete list of QI attribute values. Based on that list, the data collector or any respondent can generate the k-anonymous groups. We propose to delegate the generation of the k-anonymous groups to the data collector. There are two main reasons for this:

- *Utility*. The actual k-anonymous partition chosen may have an important impact over analyses that can be accurately performed on the k-anonymous data. The data collector is probably the one who knows best (even if often only partially) the intended use of the data and, thus, the one who can make the most appropriate partition in k-anonymous groups.

- *Performance*. Generating the k-anonymous groups is the most computationally intensive part of k-anonymity enforcement. Hence, by delegating this task to the data collector, respondents relieve themselves from this burden.

When respondents have some interest in using the anonymized data set, it is plausible to assume that any respondent will rationally collaborate to generate it. The level of protection that a respondent in a given k-anonymous group gets is dependent on the level of protection that the other respondents in the group get: as justified above in Sect. 3, k-anonymization is co-utile.

On the other side, the data collector may try to deviate from the algorithm. Since the generation of the k-anonymous partition has been delegated to the data collector, respondents must make sure before reporting confidential information that the partition computed and returned by the data collector satisfies the requirements of k-anonymity. That is, each respondent must check that her k-anonymous group comprises k or more respondents.

After verifying the partition returned by the data collector, the respondent uploads to the data collector the quasi-identifier attribute values of her k-anonymous group together with her confidential data. This communication must be done through an anonymous channel (e.g. Tor [2]) to prevent anyone (the data collector, an intruder or anyone else) from tracking the confidential data to any respondent.

The above described steps to collaboratively generate a k-anonymous data set are formalized in Protocol 1.

Protocol 1

1. *Let* $R_1, \ldots, R_m$ *be the set of respondents. Let* (qi_i, c_i) *be the quasi-identifier and confidential attribute values of* R_i*, for* $i = 1, \ldots, m$.
2. *Each* R_i *uploads her* qi_i *to a central data store so that anyone can query for* qi_i.
3. *The data collector generates a* k*-anonymous partition* $\{P_1, \ldots, P_p\}$ *and uploads it to the central data store.*
4. *Each* R_i *checks that her* k*-anonymous group* $P_{g(R_i)}$ *contains* k *or more of the original quasi-identifiers.*
   If that is not the case, R_i *refuses to provide any confidential data and exits the protocol.*
5. *Each* R_i *sends* $(P_{g(R_i)}, c_i)$ *to the data collector through an anonymous channel.*
6. *With the confidential data collected, the data collector generates the* k*-anonymous data set.*

Protocol 1 is compatible with any strategy to generate the k-anonymous partition. Possible strategies include:

- *Methods reducing the detail of the quasi-identifier attributes*. Options here are generalization and supression [14, 15, 17], or microaggregation [8]).
- *Methods breaking the connection between quasi-identifier attributes and confidential attributes*. Among these we have Anatomy [25] (that splits the data into two tables, one containing the original quasi-identifier values and the other the original confidential attribute values, with both tables being connected through a

group identifier attribute) and probabilistic k-anonymity [20] (that seeks to break the relation between quasi-identifiers and confidential attributes by means of a within-group permutation).

In fact, since the data collector and the respondents all know the exact values of the quasi-identifiers and the confidential attributes in each k-anonymous group, each of them can generate the k-anonymous data that suits her best.

In essence, the proposed protocol offers the same privacy protection as local anonymization (confidential data are only provided by the respondents in an anonymized form) while maintaining the data utility of centralized k-anonymization. At the respondents' side, there are only some minor additional communication and integrity checking costs.

We illustrate the steps of Protocol 1 for the respondents listed in the leftmost table of Fig. 1. In Step 2 each respondent uploads her quasi-identifiers. The uploaded data are shown in the center-left table of Fig. 1. At Step 3 the data collector analyzes the data uploaded in Step 2 and generates the partition in k-anonymous groups (for $k = 4$); that is, for each R_i, the data collector fixes the value of $g(R_i)$, the group assigned to R_i). This partition is shown in the center-right table of Fig. 1. For ease of understanding, the records have been arranged in a way that the k-anonymous partition P_1 contains the first k records and partition P_2 contains the last k records; that is, $P_{g(R_1)} = P_{g(R_2)} = P_{g(R_3)} = P_{g(R_4)} = P_1$ and $P_{g(R_5)} = P_{g(R_6)} = P_{g(R_7)} = P_{g(R_8)} = P_2$. In Step 4 each respondent checks that her group contains k or more of the quasi-identifier values uploaded in Step 2. Since this condition holds for all respondents in the example of the figure, respondents proceed to Step 5. In Step 5 each respondent uploads, through an anonymous channel, the group identifier she has been assigned, $P_{g(R_i)}$, together with her value for the confidential/sensitive attribute. The result is shown in the rightmost table of Fig. 1. Here the layout of the rightmost table can be misleading: although we list in the i-th row the salary of R_i for $i = 1, \ldots 4$, any permutation of the four salaries could be listed (all four salaries in the P_1 group are indistinguishable). A similar comment holds for rows 5–8, in which we could list any permutation of the salaries in the P_2 group. At this point, the

	QI		Sensitive
	Zip	Age	Salary
R_1	13053	28	35000
R_2	13068	29	30000
R_3	13068	21	20000
R_4	13053	23	27000
R_5	14853	50	40000
R_6	14853	55	43000
R_7	14850	47	48000
R_8	14850	49	45000

Step 2	
Zip	Age
13053	28
13068	29
13068	21
13053	23
14853	50
14853	55
14850	47
14850	49

	Step 3	
	Zip	Age
P_1	13053	28
	13068	29
	13068	21
	13053	23
P_2	14853	50
	14853	55
	14850	47
	14850	49

	Step 5
	Salary
P_1	35000
P_1	30000
P_1	20000
P_1	27000
P_2	40000
P_2	43000
P_2	48000
P_2	45000

Fig. 1 Distributed collaborative k-anonymization. Step numbers refer to Protocol 1. Each *row* in the *leftmost* table is only seen by the corresponding respondent. The other three tables are entirely seen by all respondents and the data collector

data collector (and the respondents) can generate the k-anonymous data set using the method they like best using that they see all tables in Fig. 1 except the leftmost one.

Distributed anonymization based on hiding in a group via manipulation of the quasi-identifiers has an important flaw. An attacker may try to simulate one or more respondents, in order to gain more insight into the k-anonymous groups. To thwart this kind of attack, we need to make sure that every respondent has a verified identity, possibly by having all respondents registered with some trusted authority. If that is not feasible, some mitigation measures can be put in place to make it more difficult for an attacker to adaptively fabricate quasi-identifier values similar to a target respondent in order to track her:

- One option is for the data store manager (maybe the data collector) to unlock the access to the quasi-identifiers list (of Step 2) only after every respondent has uploaded her quasi-identifiers. In this way, the attacker must generate his quasi-identifier values without knowing the quasi-identifier values of the other respondents. This option has the shortcoming that respondents need to trust the data store manager to perform the above access control.
- An alternative that does not require trust in any central entity is to have each respondent upload a commitment (in the cryptographic sense, [9]) to her quasi-identifiers before any actual quasi-identifier is uploaded. In this way, each respondent can check that none of the uploaded values was forged to target a specific respondent.

In the following section, we explore distributed anonymization based on masking the confidential attributes, rather than on hiding in a group via quasi-identifier manipulation.

5 Collaborative Masking of Confidential Data

Although k-anonymity is a popular privacy model, it has some important limitations. First of all, attribute disclosure is possible, even without re-identification, if the variability of the confidential attribute(s) within a k-anonymous group is small. Also, k-anonymity assumes that confidential attributes are not used in re-identification (i.e. that no confidential attribute is also a quasi-identifier), but this may not be the case if the attacker knows some confidential data. Moreover, we mentioned in the previous section that in our distributed generation of the k-anonymous data set, an attacker might simulate respondents to gain more insight into the k-anonymous groups. To deal with these issues, this section takes a different approach to generate the anonymized data set: instead of hiding within a group of respondents, each respondent masks her confidential data.

In this section, we relax the assumption that the set of quasi-identifier attributes and the set of confidential attributes are disjoint. The only assumption we make is that releasing the marginal distribution of a confidential attribute is not disclosive. What needs to be masked in the relation between a confidential attribute and any other attribute. Thus, we consider a data set with attributes $(A, C_1, \ldots, C_d)$ where

C_j are confidential attributes for $j = 1, \ldots, d$ and A groups all non-confidential attributes.

Since the marginal distribution of confidential attributes is not disclosive, respondents can share the contents of each confidential attribute among themselves and with the data collector, so that all of them get the complete list of values for each confidential attribute. In this way, each respondent can evaluate the sensitivity of her value for each confidential attribute by taking into account the values of the other respondents for that attribute. From this sensitivity evaluation, the respondent can make a more informed decision regarding the amount of masking she needs to use.

Thus, we assume that each respondent R_i makes a decision about the amount of masking required for her confidential data and reports to the data collector the tuple $(a_i, c'_{1i} \ldots, c'_{di})$, where a_i is the original value of the non-confidential attributes and c'_{1i} the masked value of confidential attribute C_i. The fact that each respondent freely and informedly decides on the amount of masking required for her confidential data is a strong privacy guarantee (the respondent can enforce the level of permutation she wishes with respect to the original values). In fact, even if the data collector or any other entity recommend a specific amount of masking, respondents are free to ignore this recommendation. For a rational respondent, the selected level of masking is based on both privacy and utility considerations.

The reported masked data can be directly used to generate the masked data set. Better yet, by applying reverse mapping [4, 16], the original marginal distribution of each confidential attribute can be recovered. This reverse mapping can be performed by the data collector and also by each respondent (because all respondents know the marginal distribution of the original attributes).

The previous discussion is formalized in Protocol 2.

Protocol 2

1. *Let $R_1, \ldots, R_m$ be the set of respondents. Let $(a_i, c_{i1} \ldots, c_{id})$ be the attribute values of R_i.*
2. *For each confidential attribute C_j, each respondent R_i uploads c_{ij} to a central data store through an anonymous channel.*
3. *For each confidential attribute C_j, each respondent R_i analyzes all attribute values and decides on the amount of masking required for c_{ij}. Let c'_{ij} be the masked value.*
4. *Each respondent R_i uploads $(a_i, c'_{i1} \ldots, c'_{id})$ to the data store.*
5. *The data collector applies reverse mapping to the data uploaded in Step 4 in order to obtain the final anonymized data set. (The same can be done by each respondent.)*

Although the reasons why Protocol 2 is safe have already been presented in the discussion prior to the algorithm formalization, a more systematic analysis is presented in the following proposition.

Proposition 8.1 *At the end of Protocol 2, nobody learns information about any respondent R_i that is more accurate than the masked data reported by R_i in Step 4.*

Proof Apart from the release of the masked data in Step 4, the only step in which R_i releases data is Step 2. Since the data released in Step 2 are not anonymized, we need to make sure they cannot be linked back to R_i.

Since the uploads in Step 2 are performed through an anonymous channel, there is no way for an attacker to track the data transfers to any particular respondent. What is more, since each c_{ij} is separately uploaded through the anonymous channel, there is no way for the attacker to link to one another the values c_{ij}, $j = 1, \ldots, d$ corresponding to the same respondent R_i (if the attacker could link such values, he could reconstruct the original record of R_i).

Finally, since by assumption releasing the marginal distribution of each confidential attribute is not disclosive, there is no risk in uploading each c_{ij} in Step 2. The reason is that each c_{ij} carries less information than the marginal distribution of attribute C_j. (The release of a c_{ij} could be problematic if C_j contains confidential information and, at the same time, can be used in re-identification, but assuming that the marginals are not disclosive rules out this situation). □

We illustrate the steps of Protocol 2 for the respondents listed in the leftmost table of of Fig. 2. We assume that Age and Salary are the confidential attributes. In Step 2 each respondent uploads to the central data store each of her values for the confidential attributes. Each respondent performs a separate upload through an anonymous channel for each of the confidential attributes. At the end of Step 2, the marginal distribution of the confidential attributes is available to the data collector and all respondents in the central data store, as illustrated in the center-left table of Fig. 2. In Step 3 each respondent can analyze the marginal distributions and decide on the amount of masking required for each confidential attribute. In this example, Age is masked by adding a random value between −5 and 5, and Salary is masked by adding a random value between −5000 and 5000. Of course, each respondent could have applied a different masking. In Step 4 each respondent uploads the masked confidential attributes together with the rest of attributes (the non-confidential ones). This upload need not be done through an anonymous channel, because all confidential data are masked. The data set uploaded by respondents to the central data store at

		Sensitive	
	Zip	Age	Salary
R_1	13053	28	35000
R_2	13068	29	30000
R_3	13068	21	20000
R_4	13053	23	27000
R_5	14853	50	40000
R_6	14853	55	43000
R_7	14850	47	48000
R_8	14850	49	45000

Step 2	
Age	Salary
28	35000
29	30000
21	20000
23	27000
50	40000
55	43000
47	48000
49	45000

Step 4		
Zip	Age	Salary
13053	29	37306
13068	24	27765
13068	18	18951
13053	19	28151
14853	51	36879
14853	50	42631
14850	52	45585
14850	49	40390

Step 5		
Zip	Age	Salary
13053	29	40000
13068	28	27000
13068	21	20000
13053	23	30000
14853	50	35000
14853	49	45000
14850	55	48000
14850	47	43000

Fig. 2 Distributed collaborative masking of the confidential attributes. Step numbers refer to Protocol 2. *Each row* in the *leftmost* table is only seen by the correspoding respondent. The other three tables are entirely seen by all respondents and the data collector

the end of Step 4 is shown in the center-right table of Fig. 2. In the final step, the data collector applies reverse mapping to each confidential attribute to recover the original marginal distributions, as illustrated in the rightmost table of Fig. 2.

6 Conclusions

We have sketched two protocols for collaborative microdata anonymization. The first one assumes a clear separation between confidential attributes and quasi-identifiers, and seeks to attain k-anonymity. In the second one, no separation between quasi-identifiers and confidential attributes is assumed, and the goal is to sufficiently mask the confidential attributes. The main difference between both methods lies on the attributes that are masked to preserve privacy. Collaborative k-anonymity masks the quasi-identifiers, thereby thwarting exact re-identification based on them. Hence, it should be preferred when we want to keep the values of the confidential attributes unmodified and limiting the probability of re-identification based on a prefixed set of quasi-identifiers is seen to be sufficient protection. Collaborative masking of confidential attributes is only concerned with masking the value of the confidential attributes in such a way that, even if re-identification happens, the intruder cannot learn with certainty the value of the confidential attributes. Hence, collaborative masking should be preferred when respondents are not comfortable with releasing fully accurate confidential data. For instance, this may be the case when the set of quasi-identifiers cannot be clearly determined, as it happens when the intruder may know some confidential pieces of information, thus potentially turning each attribute into a quasi-identifier.

Compared to local anonymization, collaborative anonymization incurs less information loss and achieves the same privacy versus the data collector. Compared to centralized anonymization, collaborative anonymization requires less trust in the data collector and achieves the same data utility. Therefore, collaborative anonymization should be preferred by rational respondents to both local and centralized anonymization.

In a statistical survey, the motivations for respondents to report data and report them truthfully to the data collector are in general unclear. As a rule, a rational respondent is willing to participate only if the benefit she obtains is greater than the potential harm due to privacy loss. If respondents are interested in the collected data set and they wish it to be as accurate as possible, then collaborative anonymization protocols are co-utile.

Acknowledgements Funding by the Templeton World Charity Foundation (grant TWCF0095/AB60 “CO-UTILITY”) is gratefully acknowledged. Also, partial support to this work has been received from the Government of Catalonia (ICREA Acadèmia Prize to J. Domingo-Ferrer and grant 2014 SGR 537), the Spanish Government (projects TIN2014-57364-C2-1-R “SmartGlacis”, TIN2015-70054-REDC and TIN2016-80250-R “Sec-MCloud”) and the European Commission (projects H2020-644024 “CLARUS” and H2020-700540 “CANVAS”). The authors are with the UNESCO Chair in Data Privacy, but the views in this work are the authors’ own and are not necessarily shared by UNESCO or any of the funding bodies.

References

1. Agrawal, S., Haritsa, J.R.: A framework for high-accuracy privacy-preserving mining. In: Proceedings of the 21st International Conference on Data Engineering, pp. 193–204 (2005)
2. Dingledine, R., Mathewson, N., Syverson, P.: Tor: the second-generation onion router. In: Proceedings of the 13th Conference on USENIX Security Symposium, pp. 21–21. CA, USA, Berkeley (2004)
3. Domingo-Ferrer, J., Martínez, S., Sánchez, D., Soria-Comas, J.: Co-utility: self-enforcing protocols for the mutual benefit of participants. Eng. Appl. Artif. Intell. **59**, 148–158 (2017)
4. Domingo-Ferrer, J., Muralidhar, K.: New directions in anonymization: permutation paradigm, verifiability by subjects and intruders, transparency to users. Inf. Sci. **337–338**, 11–24 (2015)
5. Domingo-Ferrer, J., Sánchez, D., Soria-Comas, J.: Co-utility: self-enforcing collaborative protocols with mutual help. Prog. Artif. Intell. **5**(2), 105–110 (2016)
6. Domingo-Ferrer, J., Sanchez, D., Soria-Comas, J.: Database Anonymization: Privacy Models, Data Utility, and Microaggregation-based Inter-model Connections. Morgan & Claypool, San Rafael (2016)
7. Domingo-Ferrer, J., Soria-Comas, J.: Anonymization in the time of big data. In: Privacy in Statistical Databases-PSD 2016, LNCS 9867, pp. 105–116. Springer, Berlin (2016)
8. Domingo-Ferrer, J., Torra, V.: Ordinal, continuous and heterogeneous k-anonymity through microaggregation. Data Min. Knowl. Discov. **11**(2), 195–212 (2005)
9. Goldreich, O.: Foundations of Cryptography, vol. 1, Basic Tools. Cambridge University Press, Cambridge (2001)
10. Hundepool, A., Domingo-Ferrer, J., Franconi, L., Giessing, S., Nordholt, E.S., Spicer, K., de Wolf, P.-P.: Statistical Disclosure Control. Wiley, London (2012)
11. Jiang, W., Clifton, C.: Privacy-preserving distributed k-anonymity. In: Proceedings of the 19th Annual IFIP WG 11.3 Working Conference on Data and Applications Security-DBSec 2005, LNCS 3654, pp. 166–177. Springer, Berlin (2005)
12. Jiang, W., Clifton, C.: A secure distributed framework for achieving k-anonymity. VLDB J. **15**(4), 316–333 (2006)
13. Jurczyk, P., Xiong, L.: Distributed anonymization: achieving privacy for both data subjects and data providers. In: Proceedings of the 23rd Annual IFIP WG 11.3 Working Conference on Data and Applications Security-DBSec 2009, LNCS 5645, pp. 191–207. Springer, Berlin (2009)
14. LeFevre, K., DeWitt, D.J., Ramakrishnan, R.: Incognito: efficient full-domain k-anonymity. In: Proceedings of the 2005 ACM SIGMOD International Conferenceon Management of Data, pp. 49–60. NY, USA, New York (2005)
15. LeFevre, K., DeWitt, D.J., Ramakrishnan, R.: Mondrian multidimensional k-anonymity. In: Proceedings of the 22nd International Conference on Data Engineering, Washington, DC, USA (2006)
16. Muralidhar, K., Sarathy, R., Domingo-Ferrer, J.: Reverse mapping to preserve the marginal distributions of attributes in masked microdata. In: Domingo-Ferrer, J. (ed.) Privacy in Statistical Databases, LNCS 8744, pp. 105–116. Springer, Berlin (2014)
17. Samarati, P., Sweeney, L.: Protecting privacy when disclosing information: k-anonymity and its enforcement through generalization and suppression. In: Proceedings of the IEEE Symposium on Research in Security and Privacy (1998)
18. Sánchez, D., Martínez, S., Domingo-Ferrer, J.: Comment on 'Unique in the shopping mall: on the reidentificability of credit card metadata'. Science **351**, 1274 (2016)
19. Song, C., Ge, T.: Aroma: a new data protection method with differential privacy and accurate query answering. In: Proceedings of the 23rd ACM International Conference on Conference on Information and Knowledge Management, pp. 1569–1578. NY, USA, New York (2014)
20. Soria-Comas, J., Domingo-Ferrer, J.: Probabilistic k-anonymity through microaggregation and data swapping. In: Proceedings of the IEEE International Conference on Fuzzy Systems, pp. 1–8 (2012)
21. Soria-Comas, J., Domingo-Ferrer, J.: Big data privacy: challenges to privacy principles and models. Data Sci. Eng. **1**(1), 21–28 (2015)

22. Soria-Comas, J., Domingo-Ferrer, J.: Co-utile collaborative anonymization of microdata. In: Modeling Decisions for Artificial Intelligence-MDAI 2015, LNCS 9321, pp. 192–206. Springer, Berlin (2015)
23. Wang, K., Fung, B.C.M., Dong, G.: Integrating private databases for data analysis. In: Proceedings of IEEE International Conference on Intelligence and Security Informatics, pp. 171–182. Atlanta GA (2005)
24. Warner, S.L.: Randomized response: a survey technique for eliminating evasive answer bias. J. Am. Stat. Assoc. **60**(309), 63–69 (1965)
25. Xiao, X., Tao, Y.: Anatomy: simple and effective privacy preservation. In: Proceedings of the 32rd International Conference on Very Large Data Bases, pp. 139–150 (2006)
26. Xiao, X., Tao, Y.: Personalized privacy preservation. In: Proceedings of the 2006 ACM SIGMOD International Conference on Management of Data, pp. 229–240. NY, USA, New York (2006)

Aspects of Coalitions for Environmental Protection Under Co-utility

Dritan Osmani

Abstract The game theoretic modeling of coalitions for environmental protection within the framework of a new concept of co-utility [8] is analysed. The co-utility concept can be described by two elements. Firstly, agents can improve their payoffs by collaborating with each other. Secondly, the outcome of collaboration is stable. The similarity of co-utility with common concepts of coalition stability for environmental protection is shown. But the co-utility concept is more extensive and can serve as an umbrella in all applications where agents have room for simultaneous improvements of payoffs. The development from a myopically stable outcome to a farsightedly stable outcome is discussed.

1 Introduction

The body of literature on coalitions for environmental protection (or International Environmental Agreements (IEA)) shows two conflicting views. One is rooted in cooperative game theory and concludes that the grand coalition is stable by using the core concept and by implementing transfers, as countries have different benefits and cost functions from pollution abatement [5, 6, 9]. The other view is rooted in the non-cooperative game theory, which has become the dominant path in the literature [1, 2, 4, 11, 17, 22, 24]. This paper investigates coalitions for environmental protection within the framework of *the co-utility concept* [8].

Co-utility is going to be defined based on stability and profitability. Co-utility can be described by two conditions:

1. The first condition requires that at least a part of all countries can increase the payoff (or profit) simultaneously by forming *a co-utile outcome*. The more general notion of outcome when modeling coalitions for environmental protection is a coalition structure, and a co-utile outcome is also is a special type of coalition structure.

D. Osmani (✉)
Department of Computer Engineering and Mathematics, Universitat Rovira i Virgili, Tarragona, Catalonia, Spain
e-mail: dritan.osmani@fundacio.urv.cat

J. Domingo-Ferrer and D. Sánchez (eds.), *Co-utility*, Studies in Systems, Decision and Control 110, DOI 10.1007/978-3-319-60234-9_9

2. The second condition requires that *a co-utile outcome be stable*, which means that no country or group of countries prefer to deviate from this co-utile outcome. It is obvious that this second condition is closely connected to the stability concept. As there are different stability notions, we will have different forms of co-utility notions.

Taking our previous papers [19–21, 23] as a starting point, we discuss coalitions for environmental protection within the framework of co-utility. This paper is organized as follows. Section 2 presents the game-theoretic model and considers the *single myopically and farsightedly* stable coalition co-utile outcomes. Section 3 presents a discussion of the development of a single coalition co-utile outcome from the myopic co-utility space to the farsighted co-utility space. Section 4 lists the conclusions. The Appendix presents a brief description of the FUND model, which provides the cost-benefit functions for greenhouse gas mitigation, the case study of this paper.

2 Our Model

There are 16 world regions (we name the set of all regions by N) in our game-theoretic model of IEAs (or coalitions), which are shown in the first column of Table 1. At the beginning, the link between the economic activity and the physical environment is established in order to generate the integrated assessment model. This link is established through a social welfare function calibrated to the FUND model (see Eq. (3) in the Appendix). The social welfare function captures the profit from pollution and the environmental damage. Following this approach, countries play a two-stage game. In the first stage, each country decides whether to join the coalition $C \subseteq N$ and become a signatory (or coalition member) or stay singleton and non-signatory (membership game). These decisions lead to a *co-utile outcome* or *coalition structure* S with c coalition-members and *16-c* non-members. A coalition structure fully describes how many coalitions are formed (*we are going assume that we have only one coalition*[1]), how many members each coalition has and how many singleton players there are. In the second stage, every country decides on emissions. Within the coalition, players play cooperatively (by maximizing their joint welfare) while the coalition and single countries compete in a non-cooperative way (by maximizing their own welfare).

In order to be able to compare the payoffs of the coalition members, we need to introduce the definition of partition function. Let us recall that $N = \{1, ..., 16\}$ is the set of regions (players), and that nonempty subsets of N are called coalitions. A partition[2] $\mathbf{a}$ is a set of disjoint coalitions, $\mathbf{a} = \{C_1, C_2, ..., C_k\}$, so that their union is N; the set of all partitions is $\mathscr{P}$, and the set of all the partitions where coalition C is part of, is $\mathscr{P}(C)$.

[1]For the case of multiple coalitions (or a general coalitions structure) refer to [19, 21, 23].

[2]A characteristic function approach would have been sufficient for this paper but we prefer to stick to the approach of our former papers [19, 23], whcih allows for a more general approach.

Definition 1 The partition function is a mapping $\Pi : (C, \mathscr{P}) \mapsto \Re$ where $C \in \mathscr{P}$, which assigns a value to each coalition in every partition.

Definition 2 The per-member partition function is a mapping $\pi(i)_{i \in C} : (C, \mathscr{P}) \mapsto \Re$ where $C \in \mathscr{P}$, which assigns a payoff value $\pi(i)^*_{i \in C}(C, \mathscr{P})$ to every member of each coalition in every partition.

The per-member partition function $\pi(i)_{i \in C}(C, \mathscr{P}(C))$ gives **a payoff value (or a profit)** $\pi(i)^*_{i \in C}(C, \mathscr{P}(C))$ (from now on I will write simply $\pi(i)^*_{i \in C}$) to every coalition member. This helps us to transform the preference relation to a comparison of coalition member payoffs. We would like to emphasise that the cost-benefit function (see Eq. (3)) estimated from the FUND model generates our per member partition function.

2.1 Single Myopic Co-utile Outcome

Firstly we would like to make clear that *a co-utile outcome is a coalition*, as we consider only single coalitions. As profitable coalitions (or co-utile outcomes) play a central role in our analysis, we define them from the beginning.

Definition 3 The situation in which each country maximizes its own profit, and the maximum coalition size is one is referred to as the fully non-cooperative structure.

A coalition that performs better than the fully non-cooperative structure is a *profitable coalition*. The definition of a profitable coalition is introduced below.

Definition 4 A coalition C is profitable (or individually rational) if and only if it satisfies the following condition:

$$\forall i \in C \quad \pi(i)^*_C \geq \pi(i)^*_{ind}$$

where $\pi(i)^*_C$, $\pi(i)^*_{ind}$ are the profits of country i as a member of C and in the fully non-cooperative structure, respectively.

Now we are able to state a formal definition of co-utile outcome.

Definition 5 A co-utile outcome is a profitable and stable coalition structure.

We are going to employ myopic and farsighted stability in order to characterise a co-utile outcome. Myopic stability considers only single-player moves. In other words, players are myopic if they see only one move ahead. A country that leaves (or joins) the coalition assumes that the rest of coalition members stay in the coalition and singletons (the countries that do not belong to coalition) remain non-members.

Definition 6 A coalition **C** is myopically stable (or a myopic co-utile outcome) if and only if:

- C is profitable;
- C is internally myopically stable: $\forall$ player $i \in C$, $\pi(i)^*_{i \in C} > \pi(i)^*_{C \setminus \{i\}}$;
- C is externally myopically stable: $\forall$ player $j \in C \pi(j)^*_{C \setminus \{j\}} > \pi(j)^*_{j \in C}$ or $\exists \ i \in C \mid \pi(i)^*_{C \cup \{j\}} < \pi(i)^*_{j \notin C}$,

where $\pi(i)^*$ $\pi(j)^*$ are profit function values of countries i and j.

The first condition of Definition 6 requests that the coalition C be profitable[3]; the second condition states that the coalition C is internally myopically stable only if the profit of every coalition member that leaves the coalition C decreases; the third condition states that the coalition C is externally myopically stable if, for every country that joins the coalition C, the profit decreases for the joining country or for a previous member of the coalition. In open membership games, the definition of myopic stability requests only that a country that joins a coalition reduces its profit [1]. It is more realistic (as an exclusive membership game in our case) to add the second part that a previous member of the coalition reduces its profit. In the beginning all profitable coalitions are found. Finding all profitable coalitions can be done using a simple algorithm, although at a considerable computational effort. Then, for every coalition: (i) check if all its members have higher profit compared to a fully non-cooperative structure; (ii) check internal and external myopic stability.

2.2 Farsighted Stability and Farsighted Co-utile Outcomes

It is clear that we need to consider only *profitable coalitions*.[4] Considering only profitable coalitions also reduces the computational effort required to find farsightedly stable coalitions. Before presenting our approach to finding farsighted co-utile outcomes, the definitions of *inducement process*, *credible objection*, and *farsighted stability* are presented below.

Definition 7 A coalition C_l can be induced from any coalition C_1 if and only if:

- there exist $C_1, C_2 \ldots C_{l-1}, C_l$,
  where $\pi^*_l(i) \geq \pi^*_1(i) \quad \forall i \in C_l$ and $C_q \cap C_r \neq \emptyset$ for $q, r \in [1 \ldots l]$

or

- there exist $C_1, C_2 \ldots C_{l-1}, C_l$
  where $\pi^*_m(i) \geq \pi^*_1(i) \quad \forall i \in C_1 \wedge \forall i \notin C_l$ and $C_1 \subset C_2 \subset \ldots \subset C_{l-1} \subset C_l$,

where $\pi^*_l(i)$, $\pi^*_m(i)$, $\pi^*_1(i)$ are profits, $\pi^*_1(i)$ refers to situations with C_1 and $\pi^*_l(i)$, $\pi^*_m(i)$ with C_l.

[3]Myopic stability is usually defined without the condition of profitability, but we focus on myopically stable coalitions which are profitable.

[4]If one begins by considering only profitable coalitions, it is sufficient to find all profitable or non-profitable single farsightedly stable coalitions, see Observations 3.2 and 3.3 in [21].

The first part of the inducement definition requires that all countries of the coalition C_l do not decrease their profits and indirectly assumes that those countries have started the formation of the final coalition. The second part of the definition requires that all countries that leave the initial coalition C_1 (including free-riding) do not decrease their profits. The definitions of objection and credible objection are presented below:

Definition 8 A group of countries G_1 has an objection to a coalition C_1 if there exists another coalition C_n such that C_n can be induced from C_1 and $\pi_n^*(i) \geq \pi_1^*(i) \quad \forall i \in G_1$, where $\pi_n^*(i)$, $\pi_1^*(i)$ are profits, $\pi_1^*(i)$ refers to situations with C_1 and $\pi_n^*(i)$ with C_n.

Definition 9 A group of countries G_1 has a credible objection to a coalition C_1 if there exists another coalition C_n such that:

- C_n can be induced from C_1 and $\pi_n^*(i) \geq \pi_1^*(i) \quad \forall i \in G_1$;
- no group of countries G_n has a objection to C_n,

where $\pi_n^*(i)$, $\pi_1^*(i)$ are profits, $\pi_1^*(i)$ refers to situations with C_1 and $\pi_n^*(i)$ with C_n.

A group of countries has a credible objection to a coalition if there exists an inducement process, and moreover this particular inducement process is the final one (or no group of countries has any objection to the final coalition). Now, it is possible to state the definition of farsighted stability.

Definition 10 A coalition C is **farsightedly stable** if no group of countries G has a credible objection to C.

The definition of farsighted stability is based on the definition of the inducement process (no credible objection signifies that the inducement process has come to an end). This means that one needs to trace the inducement process in order to test whether a coalition is farsightedly stable or not. Definition 7 makes it clear that there are two main types of inducement process. In the first type, there is a change sequence of coalitions where *the countries in the final coalition* do not decrease their profits. In the second type, there is a change sequence of coalitions where *the countries that leave the initial coalition* do not decrease their profit. There are five classes of inducement process. Three of them belong to the first type of inducement process; the coalition grows bigger; gets smaller; some coalition-members leave the coalition and some others join it. The last two classes of inducement process belong to the second type of inducement process. The fourth class is a special one, namely *free-riding*. One or more countries leaves the coalition and increase their welfare. The fifth inducement process is also a special inducement process which occurs only in *non-profitable coalitions* that have at least one country that has a welfare smaller than in the fully non-cooperative structure. Those countries are going to leave the coalition (and increase their welfare) *not due to free-riding* but because the joint welfare is distributed unfairly; there is no credible objection against those countries. Even if the coalition is dissolved and a fully non-cooperative structure is reached, their welfare is higher than in the initial non-profitable coalition.

3 From a Myopic Co-utile Outcome to a Farsighted Co-utile Outcome: A Discussion

This section provides a description of the way from the single coalition myopic co-utile outcome (or simply myopically stable coalition) (USA, CHI, NAF) to the farsighted co-utile outcome $(USA, LAM, SEA, CHI, NAF, SSA)$. There is no free-riding initiative in the coalition (USA, CHI, NAF), as the coalition is internally myopically stable. This indicates that if a country leaves the coalition, it decreases its profit. Therefore, a myopic coalition formation ends when free-riding appears. On the opposite, the farsightedly stable coalition formation does not end when free-riding appears, but it ends when the profitability condition is not satisfied any further. As a consequence, one can design the following scheme for portraying a way from a myopic coalition to a farsightedly stable coalition:

$$(USA, CHI) \Rightarrow \underbrace{(USA, CHI, NAF)^{ds}}_{myopic\,stable\,coalition} \Rightarrow \underbrace{(USA, CHI, NAF, SSA)}_{free-riding\,appears} \Rightarrow \ldots$$

$$\Rightarrow \underbrace{(USA, LAM, SEA, CHI, NAF, SSA)^{fs}.}_{profitability\,condition\,can\,not\,be\,satisfied\,further}$$

Nevertheless, this is better seen in Fig. 1. Along the y-axis we have the profit of each country in billions of dollars. Along the x-axis, there are some possible coalitions from a fully non-cooperative structure (a_{FNS}) to a myopically stable coalition (USA, CHI, NAF), and it ends with a farsightedly stable coalition $(USA, CHI, NAF, SSA, SEA, LAM)$. When $x = 1$, we have a_{FNS}; when $x = 2$, we have (USA, CHI); when $x = 3$, we have (USA, CHI, NAF); when $x = 4$, we have (USA, CHI, NAF, SSA); when $x = 5$, we have $(USA, CHI, NAF, SSA, SEA)$; and when $x = 6$, we have $(USA, CHI, NAF, SSA, SEA, LAM)$. Every line denotes the variation in welfare (profit) of the respective country when different coalitions are formed. At the beginning of the dotted line, the respective country is not a member of a coalition. At the end of the dotted line, the respective country joins the coalition. Countries that join the coalition improve their profits until the myopic stable coalition (USA, CHI, NAF) is formed. The formation of coalition (USA, CHI, NAF) signifies the end of the myopic co-utility space. After the coalition (USA, CHI, NAF) is formed, every country that joins this coalition reduces its profit. This shows that the free-riding initiative exists, as these coalitions (all coalitions that contain the coalition (USA, CHI, NAF) as a sub-coalition) are not internally myopically stable. The formation of coalition $(USA, CHI, NAF, SSA, SEA, LAM)$ signifies the end of the farsighted co-utility space. For any enlargement, after the farsightedly stable coalition $(USA, CHI, NAF, SSA, SEA, LAM)$ is formed, the profitability condition is not satisfied any longer. This situation indicates that there is no bigger farsightedly stable coalition as *farsighted stability is a function of profitability*, which is problematical to be realised for a single large coalition: the asymmetry of countries does not permit large profitable coalitions. This is a typical situation in a myopic coalition formation,

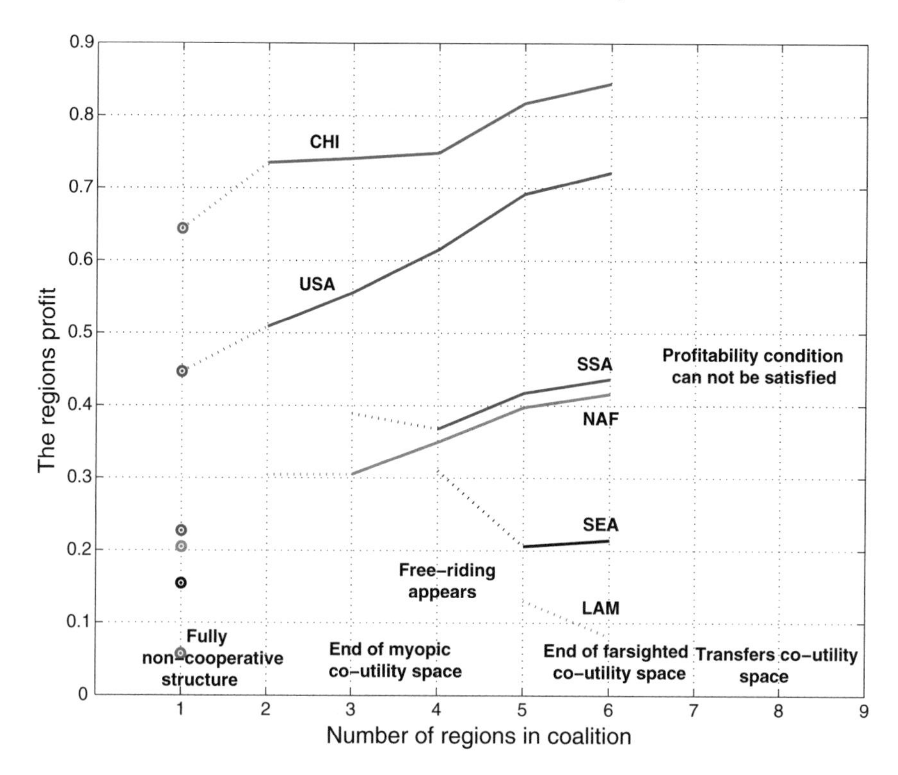

Fig. 1 Development of a single coalition co-utile outcome from myopic co-utile space to farsighted co-utile space. *Source* [20, 23]

and shows that myopic stable coalitions are frequently subsets of farsightedly stable coalitions. Moreover, *one can see the farsightedly stable coalition as the "maximum" that can be achieved by in a co-utile outcome without side payments.*

4 Conclusions

This paper has examined coalitions for environmental protection within the framework of co-utility [8]. The FUND model provides the cost-benefit functions of pollution abatement. The structure and dynamics of the damage-cost functions of the FUND model influence the results.

The myopic stability concept presumes that the players are myopic and that the formation of the co-utile outcome consists of only single-player moves. The farsighted stability takes into account the farsightedness of the players; this indicates that every country that takes deviations into account realises that a deviation may trigger further deviations that can worsen its initial position.

The largest single farsightedly stable coalition and myopically stable coalition are small. Myopic stability implies that free-riding makes it difficult to have large single stable coalitions. By contrast, farsighted stability implies that, due to asymmetry, the profitability condition is problematic to be satisfied for large single farsightedly stable coalitions. We demonstrate that myopically stable coalitions are often sub-coalitions of farsightedly stable coalitions. Moreover, farsightedly stable coalitions can be frequently the largest stable coalitions that game theory without side payments can realize. In real-world coalition formation, such as the Kyoto protocol, it is more reasonable to suppose that a part of the countries is myopic and that another part is farsighted. As a result, we should expect coalitions that are bigger than myopically stable but smaller than farsightedly stable coalitions. It will be interesting to consider more detailed regions and a game-theoretic approach that considers side payments.

Acknowledgements Support from the Templeton World Charity Foundation (grant TWCF0095/AB60 "CO-UTILITY") is gratefully acknowledged. The views in this paper are the author's own and do not necessarily reflect those of the Templeton World Charity Foundation.

Appendix. The FUND Model

This paper uses version 2.8 of the Climate Framework for Uncertainty, Negotiation and Distribution (FUND). Version 2.8 of FUND corresponds to version 1.6, described and applied by [28–30, 33], except for the impact module, which is described by [31, 32] and updated by [15]. A further difference is that the current version of the model distinguishes 16 instead of 9 regions. Finally, the model considers emission reduction of methane and nitrous oxide as well as carbon dioxide, as described by [34].

Essentially, FUND consists of a set of exogenous scenarios and endogenous perturbations. The model distinguishes 16 major regions of the world, viz. the United States of America (USA), Canada (CAN), Western Europe (WEU), Japan and South Korea (JPK), Australia and New Zealand (ANZ), Central and Eastern Europe (EEU), the former Soviet Union (FSU), the Middle East (MDE), Central America (CAM), South America (LAM), South Asia (SAS), Southeast Asia (SEA), China (CHI), North Africa (NAF), Sub-Saharan Africa (SSA), and Small Island States (SIS). The model runs from 1950 to 2300 in time steps of 1 year. The primary reason for starting in 1950 is to initialize the climate change impact module. In FUND, the impacts of climate change are assumed to depend on the impact in the previous year, in this way reflecting the process of adjustment to climate change. Because the initial values to be used for the year 1950 cannot be approximated very well, both physical and monetized impacts of climate change tend to be poorly represented in the first few decades of the model runs. The period of 1950–1990 is used for the calibration of the model, which is based on the IMAGE 100-year database [3]. The period 1990–2000 is based on observations of the World Resources Databases [35]. The climate scenarios for the period 2010–2100 are based on the EMF14 Standardized Scenario, which lies somewhere in between IS92a and IS92f [14]. The 2000–2010 period is interpolated from the immediate past, and the period 2100–2300 extrapolated.

The scenarios are defined by the rates of population growth, economic growth, autonomous energy efficiency improvements as well as the rate of decarbonization of the energy use (autonomous carbon efficiency improvements), and emissions of carbon dioxide from land use change, methane and nitrous oxide. The scenarios of economic and population growth are perturbed by the impact of climatic change. Population decreases with increasing climate change related deaths that result from changes in heat stress, cold stress, malaria, and tropical cyclones. Heat and cold stress are assumed to have an effect only on the elderly, non-reproductive population. In contrast, the other sources of mortality also affect the number of births. Heat stress only affects the urban population. The share of the urban population among the total population is based on the World Resources Databases [35]. It is extrapolated based on the statistical relationship between urbanization and per-capita income, which are estimated from a cross-section of countries in 1995. Climate-induced migration between the regions of the world also causes the population sizes to change. Immigrants are assumed to assimilate immediately and completely with the respective host population.

The market impacts are dead-weight losses to the economy. Consumption and investment are reduced without changing the savings rate. As a result, climate change reduces long-term economic growth, although consumption is particularly affected in the short-term. Economic growth is also reduced by carbon dioxide abatement measures. The energy intensity of the economy and the carbon intensity of the energy supply autonomously decrease over time. This process can be accelerated by abatement policies, an option not considered in this paper.

The endogenous parts of FUND consist of the atmospheric concentrations of carbon dioxide, methane and nitrous oxide, the global mean surface temperature, the impact of carbon dioxide emission reductions on the economy and on emissions, and the impact of the damages to the economy and the population caused by climate change. Methane and nitrous oxide are taken up in the atmosphere, and then geometrically depleted. The atmospheric concentration of carbon dioxide, measured in parts per million, is represented by the five-box model of [16]. Its parameters are taken from [12]. The model also contains sulphur emissions [34].

The radiative forcing of carbon dioxide, methane, nitrous oxide and sulphur aerosols is determined based on [25]. The global mean temperature T is governed by a geometric build-up to its equilibrium (determined by the radiative forcing RF), with a half-life of 50 years. In the base case, the global mean temperature rises in equilibrium by 2.5°C for a doubling of carbon dioxide equivalents. Regional temperature follows from multiplying the global mean temperature by a fixed factor, which corresponds to the spatial climate change pattern averaged over 14 GCMs [18]. The global mean sea level is also governed by a geometric build-up, with its equilibrium level determined by the temperature and a half-life of 50 years. Both temperature and sea level are calibrated to correspond to the best guess temperature and sea level for the IS92a scenario of [13].

The climate impact module, based on [32, 33] includes the following categories: agriculture, forestry, sea level rise, cardiovascular and respiratory disorders related to cold and heat stress, malaria, dengue fever, schistosomiasis, diarrhoea, energy consumption, water resources, and unmanaged ecosystems. Climate change related damages can be attributed to either the rate of change (benchmarked at 0.04°C) or the level of change (benchmarked at 1.0°C). Damages from the rate of temperature change slowly fade, reflecting adaptation [33]. People can die prematurely due to temperature stress or vector-borne diseases, or they can migrate because of sea level rise. Like all impacts of climate change, these effects are monetized. The value of a statistical life is set to be 200 times the annual per capita income. The resulting value of a statistical life lies in the middle of the observed range of values in the literature [7]. The value of emigration is set to be 3 times the per capita income [26, 27], the value of immigration is 40% of the per capita income in the host region [7]. Losses of dryland and wetlands due to sea level rise are modelled explicitly. The monetary value of a loss of one square kilometre of dryland was on average $4 million in OECD countries in 1990 [10]. Dryland value is assumed to be proportional to GDP per square kilometre. Wetland losses are valued at $2 million per square kilometre on average in the OECD in 1990 [10]. The wetland value is assumed to have a logistic relation to per capita income. Coastal protection is based on cost-benefit analysis, including the value of additional wetland lost due to the construction of dikes and subsequent coastal squeeze.

Other impact categories, such as agriculture, forestry, energy, water, and ecosystems, are directly expressed in monetary values without an intermediate layer of impacts measured in their 'natural' units [32]. Impacts of climate change on energy consumption, agriculture, and cardiovascular and respiratory diseases explicitly recognize that there is a climatic optimum, which is determined by a variety of factors, including plant physiology and the behaviour of farmers. Impacts are positive or negative depending on whether the actual climate conditions are moving closer to or away from that optimum climate. Impacts are larger if the initial climate conditions are further away from the optimum climate. The optimum climate is of importance with regard to the potential impacts. The actual impacts lag behind the potential impacts, depending on the speed of adaptation. The impacts of not being fully adapted to new climate conditions are always negative [33]. The impacts of climate change on coastal zones, forestry, unmanaged ecosystems, water resources, diarrhoea, malaria, dengue fever, and schistosomiasis are modelled as simple power functions. Impacts are either negative or positive, and they do not change sign [33]. Vulnerability to climate change changes with population growth, economic growth, and technological progress. Some systems are expected to become more vulnerable, such as water resources (with population growth), heat-related disorders (with urbanization), and ecosystems and health (with higher per capita incomes). Other systems are projected to become less vulnerable, such as energy consumption (with technological progress), agriculture (with economic growth) and vector- and water-borne diseases (with improved health care) [33].

Note that we make use of data only for the year 2005. This is sufficient as static game theory is used but with a sophisticated stability concept.

The Welfare Function of the FUND Model

We approximate the FUND model with a linear benefit/quadratic cost structure for the analysis of coalition formation. Specifically, the abatement cost function is represented as:

$$C_i = \alpha_i R_i^2 Y_i, \tag{1}$$

where C denotes abatement cost, R relative emission reduction, Y gross domestic product, index i denotes regions and α is the cost parameter. The benefit function is approximated as:

$$B_i = \beta_i \sum_j^n R_j E_j \tag{2}$$

where B denotes benefit, β the marginal damage costs of carbon dioxide emissions and E unabated emissions. Table 1 gives the parameters of Eqs. (1) and (2) as estimated by FUND. Moreover, the profit π_i of a country i is given as:

Table 1 Our data from the year 2005, where α is the abatement cost parameter (unitless), β the marginal damage costs of carbon dioxide emissions (in dollars per tonne of carbon), E the carbon dioxide emissions (in billion metric tonnes of carbon) and Y the gross domestic product, in billions US dollars. *Source* FUND

	α	β	E	Y
USA	0.01515466	2.19648488	1.647	10399
CAN	0.01516751	0.09315600	0.124	807
WEU	0.01568000	3.15719404	0.762	12575
JPK	0.01562780	−1.42089104	0.525	8528
ANZ	0.01510650	−0.05143806	0.079	446
EEU	0.01465218	0.10131831	0.177	407
FSU	0.01381774	1.27242378	0.811	629
MDE	0.01434659	0.04737632	0.424	614
CAM	0.01486421	0.06652486	0.115	388
LAM	0.01513700	0.26839935	0.223	1351
SAS	0.01436564	0.35566631	0.559	831
SEA	0.01484894	0.73159104	0.334	1094
CHI	0.01444354	4.35686225	1.431	2376
NAF	0.01459959	0.96627119	0.101	213
SSA	0.01459184	1.07375825	0.145	302
SIS	0.01434621	0.05549814	0.038	55

$$\pi_i = B_i - C_i = \beta_i \sum_j^n R_j E_j - \alpha_i R_i^2 Y_i. \tag{3}$$

The second derivative of $d^2\pi_i/dR_i^2 = -2\alpha_i < 0$ as $\alpha_i > 0$. It follows that the profit function of every country i is strictly concave, and as a consequence has a unique maximum. Hence, the non-cooperative optimal emission reduction is found from the first-order optimal condition:

$$d\pi_i/dR_i = \beta_i E_i - 2\alpha_i R_i Y_i = 0 \Rightarrow R_i = \beta_i E_i/(2\alpha_i Y_i). \tag{4}$$

If a region i is in a coalition with a region j, the optimal emission reduction is given by:

$$d\pi_{i+j}/dR_i = 0 \Rightarrow E_i(\beta_i + \beta_j) - 2\alpha_i R_i Y_i = 0 \Rightarrow R_i = (\beta_i + \beta_j)E_i/(2\alpha_i Y_i). \tag{5}$$

Thus, the price for entering a coalition is higher emission abatement at home. The return is that the coalition partners also raise their abatement efforts.

Note that our welfare functions are orthogonal. This indicates that the emissions change of a country does not affect the marginal benefits of other countries (that is the independence assumption). In our game, countries outside the coalition benefit from the reduction in emissions achieved by the cooperating countries, but they cannot affect the benefits derived by the members of the coalition. As our cost-benefit functions are orthogonal, our approach does not capture the effects of emissions leakage. Even so, our cost benefit functions are sufficiently realistic as they are an approximation of the complex model FUND and our procedure of dealing with farsighted stability is also general and appropriate for non-orthogonal functions.

References

1. Barrett, S.: Self-enforcing international environmental agreements. Oxf. Econ. Pap. **46**, 878–894 (1994)
2. Barrett, S.: Environment and Statecraft: The Strategy of Environmental Treaty-Making. Oxford University Press, Oxford (2003)
3. Batjes, J, Goldewijk, C.: The IMAGE 2 Hundred Year (1890–1990) Database of the Global Environment (HYDE). Report No. 410100082, RIVM, Bilthoven (1994)
4. Botteon, M, Carraro, C.: Environmental coalitions with heterogeneous countries: burden-sharing and carbon leakage. In: Ulph, A. (ed.) Environmental Policy, International Agreements, and International Trade. Oxford University Press, Oxford (2001)
5. Chander, P.: The gamma-core and coalition formation. Int. J. Game Theory **35**(4), 379–401 (2007)
6. Chander, P., Tulkens, H.: The core of an economy with multilateral environmental externalities. Int. J. Game Theory **26**(3), 379–401 (1997)
7. Cline, W.: Econ. Glob. Warm. Institute for International Economics, Washington, DC (1992)
8. Domingo-Ferrer, J., Martínez, S., Sánchez, D., Soria-Comas, J.: Co-utility: self-enforcing protocols for the mutual benefit of participants. Eng. Appl. Artif. Intell. **59**, 148–158 (2017)
9. Eyckmans, J., Tulkens, H.: Simulating coalitionally stable burden sharing agreements for the climate change problem. Resour. Energy Econ. **25**, 299–327 (2003)

10. Fankhauser, S.: Protection vs retreat: the economic costs of sea level rise. Environ. Plan. A **27**, 299–319 (1994)
11. Finus, M., van Ierland, E., Dellink, R.: Stability of climate coalitions in a cartel formation game. Econ. Gov. **7**, 271–291 (2006)
12. Hammitt, J.K., Lempert, R.J., Schlesinger, M.E.: A sequential-decision strategy for abating climate change. Nature **357**, 315–318 (1992)
13. Kattenberg, A., Giorgi, F., Grassl, H., Meehl, G.A., Mitchell, J.F.B., Stouffer, R.J., Tokioka, T., Weaver, A.J., Wigley, T.M.L.: Climate models - projections of future climate. In: Houghton, J.T., et al. (eds.) Climate Change 1995: The Science of Climate Change. Contribution of Working Group I to the Second Assessment Report of the Intergovernmental Panel on Climate Change, Cambridge University Press, Cambridge (1996)
14. Leggett, J., Pepper, W.J., Swart, R.: Emissions scenarios for the IPCC: an update. In: Houghton, J.T., Callander, B.A., Varney, S.K. (eds.) Climate Change 1992: The Supplementary Report to the IPCC Scientific Assessment, vol. 1. Cambridge University Press, Cambridge (1992)
15. Link, P.M., Tol, R.S.J.: Possible economic impacts of a shutdown of the thermohaline circulation: an application of FUND. Port. Econ. J. **3**, 99–114 (2004)
16. Maier-Reimer, E., Hasselmann, K.: Transport and storage of carbon dioxide in the ocean: an inorganic ocean circulation carbon cycle model. Clim. Dyn. **2**, 63–90 (1987)
17. McGinty, M.: International environmental agreements among asymmetric nations. Oxf. Econ. Pap. **59**, 45–62 (2007)
18. Mendelsohn, R., Morrison, W., Schlesinger, M.E., Andronova, N.G.: Country-specific market impacts of climate change. Clim. Chang. **45**, 553–569 (2000)
19. Osmani, D.: A note on computational aspects of farsighted coalitional stability (revised version). Working Paper, FNU-194 (2015)
20. Osmani, D.: Computational and conceptual aspects of coalition for environmental protection within co-utility framework. Working Paper, Co-Utility Project. Templeton World Charity Foundation, Grant TWCF0095 /AB60 CO-UTILITY (2016)
21. Osmani, D., Tol, R.S.J.: Towards farsightedly stable international environmental agreements. J. Public Econ. Theory **11**(3), 455–492 (2009)
22. Osmani, D., Tol, R.S.J.: The case of two self-enforcing international environmental agreements for environmental protection with asymmetric countries. Comput. Econ. **36**(2), 93–119 (2010)
23. Osmani, D., Tol, R.S.J.: Towards farsightedly stable international environmental agreements: part two (revised version). Working Paper, FNU-149 (2015)
24. Rubio, J.S., Ulph, U.: Self-enforcing international environmental agreements revisited. Oxf. Econ. Pap. **58**, 223–263 (2006)
25. Shine, K.P., Derwent, R.G., Wuebbles, D.J., Morcrette, J.J.: Radiative forcing of climate in climate change. In: Houghton, J.T., Jenkins, G.J., Ephraums, J.J. (eds.) The IPCC Scientific Assessment, vol. 1. Cambridge University Press, Cambridge (1990)
26. Tol, R.S.J.: The damage costs of climate change toward more comprehensive calculations. Environ. Res. Econ. **5**, 353–374 (1995)
27. Tol, R.S.J.: The damage costs of climate change towards a dynamic representation. Ecol. Econ. **19**, 67–90 (1996)
28. Tol, R.S.J.: Kyoto, efficiency, and cost-effectiveness: an application of FUND, In: A Multi-Model Evaluation, Energy Journal Special Issue on the Costs of the Kyoto Protocol, pp. 130–156 (1999a)
29. Tol, R.S.J.: Spatial and temporal efficiency in climate change: an application of FUND. Environ. Res. Econ. **58**(1), 33–49 (1999b)
30. Tol, R.S.J.: Equitable cost-benefit analysis of climate change. Ecol. Econ. **36**(1), 71–85 (2001)
31. Tol, R.S.J.: Estimates of the damage costs of climate change - part 1: benchmark estimates. Environ. Res. Econ. **21**, 47–73 (2002a)
32. Tol, R.S.J.: Estimates of the damage costs of climate change - part 2: benchmark estimates. Environ. Res. Econ. **21**, 135–160 (2002b)
33. Tol, R.S.J.: Welfare specifications and optimal control of climate change: an application of FUND. Energy Econ. **24**, 367–376 (2002c)

34. Tol, R.S.J.: Multi-gas emission reduction for climate change policy: an application of FUND. In: Energy Journal in Volume: Multi-Greenhouse Gas Mitigation and Climate Policy. Special Issue 3 (2006)
35. W.R.I. World Resources Database.: World Resources Institute, Washington, DC (2000–2001)

United We Stand: Exploring the Notion of Cooperation Support System for Business Model Design

Riccardo Bonazzi and Vincent Grèzes

Abstract In this chapter, we identify possible connections between the notion of co-utility and new business models, which are based on cooperation among consumers. We start with known cases of sharing economy (Uber and AirBnB) to introduce the business model canvas as a tool to support the design of right incentives for cooperation among stakeholders. Then we use the business model canvas to illustrate two research projects and two real-life applications focused on a transportation firm and a boutique hotel.

1 Introduction

This chapter tries to address a managerial question revolving around the new concept of co-utility [20], which is fairly straightforward: who is going to financially support a system that follows a co-utile protocol?

Such question is associated with the capacity of such system to generate and to capture value, as perceived by its stakeholders. In other words, if for example we have an algorithm that will be successfully implemented as part of an open source code, how is the system going to find the required resources to pay for the maintenance of the technical infrastructure, the manpower needed to let the code evolve and to manage the community of users? Hence, the focus of our analysis is the overall *information systems*, which we will call cooperation support systems. They refer to technological as well as organizational subsystems, where there is an interest in problems and solutions that emerge in the interactions between the technological and the organizational levels [25].

R. Bonazzi (✉) · V. Grèzes
Institute of Entrepreneurship and Management, University of Applied Sciences Western Switzerland, Maison de l'Entrepreneuriat, Techno-Pôle 3, 3960 Sierre, Switzerland
e-mail: Riccardo.Bonazzi@hevs.ch

V. Grèzes
e-mail: Vincent.Grezes@hevs.ch

J. Domingo-Ferrer and D. Sánchez (eds.), *Co-utility*, Studies in Systems, Decision and Control 110, DOI 10.1007/978-3-319-60234-9_10

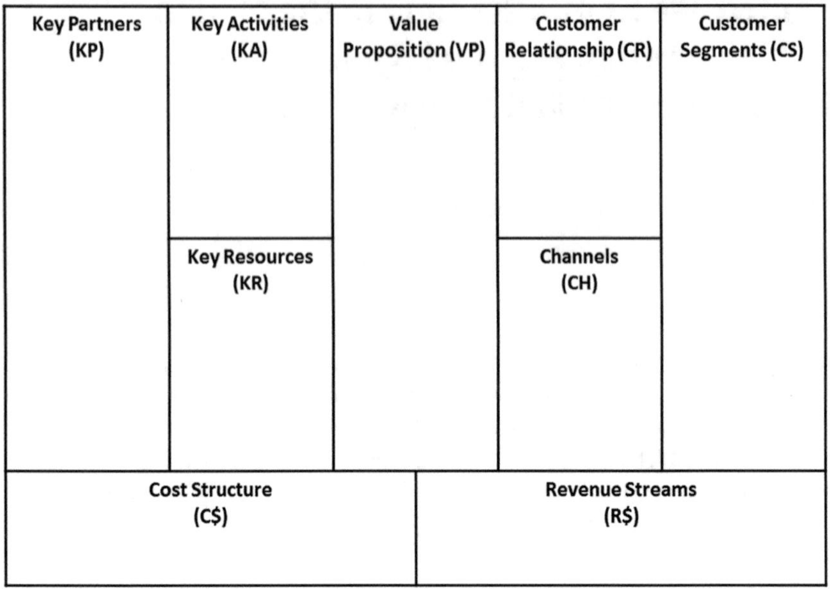

Key Partners (KP)	Key Activities (KA)	Value Proposition (VP)	Customer Relationship (CR)	Customer Segments (CS)
	Key Resources (KR)		Channels (CH)	
Cost Structure (C$)			Revenue Streams (R$)	

Fig. 1 The business model canvas [31]

Since we wish to address a practical problem, in this chapter we follow the guidelines of design science research, seeking for utility rather than truth [23]. Design theories focus on "how to do something" and give explicit prescriptions on how to design and develop technological products or managerial interventions. Indeed, [21] shows how design theory can be seen as the fifth in five classes of theories that are relevant to information systems: (1) theory for analyzing, (2) theory for explaining, (3) theory for predicting, (4) theory for explaining and predicting, and (5) theory for design and action.

In order to address our question, we need to design a business model for our information system. Among the large set of different approaches to perform business model design, which have been developed by scholars in business model innovation, a simple way to represent the business model of a company is to use the business model canvas (BMC) described by [31]. Such model, shown in Fig. 1, is composed of nine building blocks, which belong to three main parts:

(a) the customer side, which includes the customer segments, the channel and the customer relationships;
(b) the activities side, which includes the key activities, the key resources and the key partners;
(c) the revenues side, which includes the revenue streams and the cost structure.

2 A Practical Example in Four Steps: A Trip to Switzerland

To illustrate a set of different business applications built around the notion of co-utility, let us follow an imaginary couple (Alice and Bob) that is going to spend a weekend in Switzerland. We shall follow four steps of the customers' journey: (a) starting from their arrival to the airport of Geneva, (b) moving to their trip towards the hotel, (c) observing the first minutes of their check-in at the hotel and (d) thinking about their first meal.

The four cases are meant to illustrate two relevant dimensions concerning the potential for scalability for co-utile protocols, in business terms: (1) the number of users involved, since it is more complex to continuously mobilize a crowd of users instead of a small group of customers and (2) the decentralization of the information system, since a fully decentralized architecture requires a more sophisticated revenue model to be profitable.

As shown in Table 1, each case refers to a study done by the Institute of Entrepreneurship and Management of the University of Applied Sciences Western Switzerland. We limited the choice of our sources to those where we had first-hand knowledge, even though that implied having to leave aside a large set of research done by colleagues in other universities. Therefore, these cases are meant to be shown as illustration and they are not meant to be used as a comprehensive overview of the current state of the art.

To allow the comparison among the four cases, each of them is divided into three paragraphs:

1. *Story*: the first paragraph shows how the story of Alice and Bob evolves and it briefly explains how the situation could be improved by a co-utile protocol.
2. *Theoretical model and test results*: the second paragraph illustrates the elements of our theory and shares outcomes obtained from a large set of testing techniques (survey, action research, simulation).
3. *Business model insights*: the last paragraph uses the business model canvas to illustrate why the case can be considered as interesting to further expand our understanding of cooperation support systems. To facilitate the overall understanding, the business model of each case builds on the notion explored in the previous cases and the images of the three canvases are presented in a cumulative fashion. In this way, we move from the simplest case to the most complex one, as indicated by the numbers of Table 1.

Table 1 The four cases and their references

	Partially decentralized	Fully decentralized
Group of users	(3) Case study: maya boutique hotel [10]	(1) Research project: privacy for security [26]
Crowd	(2) Case study: tooxme [8]	(4) Research project: healthy lottery [6]

3 Step A: Arrival at Geneva Airport and Connection to the Wi-Fi. Introducing the Notion of Privacy for Security

Let us imagine that, as soon as they arrive to Geneva airport, Alice and Bob turn on their wi-fi connections to check their e-mails and to let their families know that they arrived in Switzerland. In order to access internet, Alice and Bob have to fill in a form with their personal data and they will be sending data about password and personal information via a wi-fi connection, whose degree of protection is sometime fairly limited.

A large majority of research papers in the field of information security assumes that Alice and Bob are unaware of the risks they are exposed to while they are connected to the wi-fi. Nonetheless, some scholars in the field of security economics believe that Alice and Bob estimate that the risk they are taking is outgrown by the payoff.

The success of the privacy management solution relies on the development of technology and regulations to protect personal information [15]. Privacy is a dynamic and dialectic process of give and take between and among technical and social entities in ever-present and natural tension with the simultaneous need for information to be made public [33]. We therefore understand the mobile user and the service provider as both competing and cooperating to gain access to a valuable resource (mobile user's data) as described by [28]. In this model, a customer can decide to protect the personal data or not, and if the customer cares about the protection of personal data, the firm can decide rather to protect the customer's data or not.

Accordingly, we start with two considerations about privacy management, adapted from [1]: (a) as long as firms consider privacy as a cost, they will try to minimize the required investments, (b) rationally bounded users are willing to sacrifice security for ease of use. Therefore, we shall assess how to design an information system that assures profitability for a firm, while addressing the needs of a group of customers.

3.1 Theoretical Model and Results of a Survey Study

Our theoretical model has four constructs and three hypotheses. Since previous studies have already focused on the effects of antecedents, we focus on the effects among three of these antecedents, which will be explained in the rest of the paragraph. Accordingly, in our model we define user payoff as the degree to which a mobile user perceives as fair the benefits he or she receives from mobile service companies in return for providing personal information; and we define *personal data disclosed* as the degree to which a mobile user perceives whether personal data is disclosed by the mobile service company. Thus, we claim that the user payoff is found to be one major predictor for personal data disclosed and we state that *(H1): The personal data disclosed has a negative effect on user payoff.*

In exchange for user data, the m-commerce provider could offer a service, which is either standard or fully customized. We introduce the concept of *service personalization*, which we define as the degree of fairness that a mobile user perceives relative to mobile service company's treatment of information privacy. Service personalization depends on customer willingness to share information and use personalized services [14]. It is natural to expect a positive relationship between the amount of personalization available and users' benefit. Hence we state that *(H2): The amount of personalization available has (a) a direct and positive effect and (b) interacts with the effect between personal data disclosed and user payoff (in other words it has a moderating effect).*

As consumers take relatively high risks by submitting personal data to the mobile service provider, data controls over user personal data using privacy metrics are a useful tool to decrease user concern for privacy risks [22]. Lack of such controls decreases mobile user trust in the provider and lowers the perceived payoff [16]. Hence we state that *(H3): The amount of control over user personal data has (a) a direct and positive effect and (b) a moderating effect on the relationship between personal data disclosed and user payoff.*

The extent of data disclosed always has a significant effect on user payoff ($p < 0.01$ in all four steps), which is negative (−2.004 in the first step). In other words, it appears that mobile users sacrifice certain benefit or increase their concerns for risk when the service asks for their personal information. Thus, H1 is strongly supported.

Recalling [27], we expect people who show generally low risk aversion to have different opinions on their payoffs as opposed to those who are highly risk-averse. Thus, we divide our sample into two clusters accordingly. We adopt the median cluster method based on two variables: subjects' global concerns for privacy and concerns in the mobile service sector. We conduct regression analysis for both clusters, and we find that for people who have a relatively high level of concern about privacy when providing personal information (risk-averse users), neither personalization available nor user control over personal data play an important role in determining payoff. This interpretation extends previous analyses on why privacy policies of websites are often not shown in the first page [11]. For people who have a relatively low level of concern about privacy when providing personal information (risk-neutral users), both variables are demonstrated to be essential indicators. Hence, H2 and H3 are rejected for risk-averse mobile users.

However, Fig. 2 shows that further analysis reveals significant effects of personalization available and user's control for risk-neutral users. In particular, personalization being available has a statistically significant effect on the user's payoff (0.912, $p < 0.05$), and user's control over personal data also has a statistically significant impact on the user's payoff (1.132, $p < 0.01$). Thus, for risk-neutral mobile users, H2 and H3a are supported.

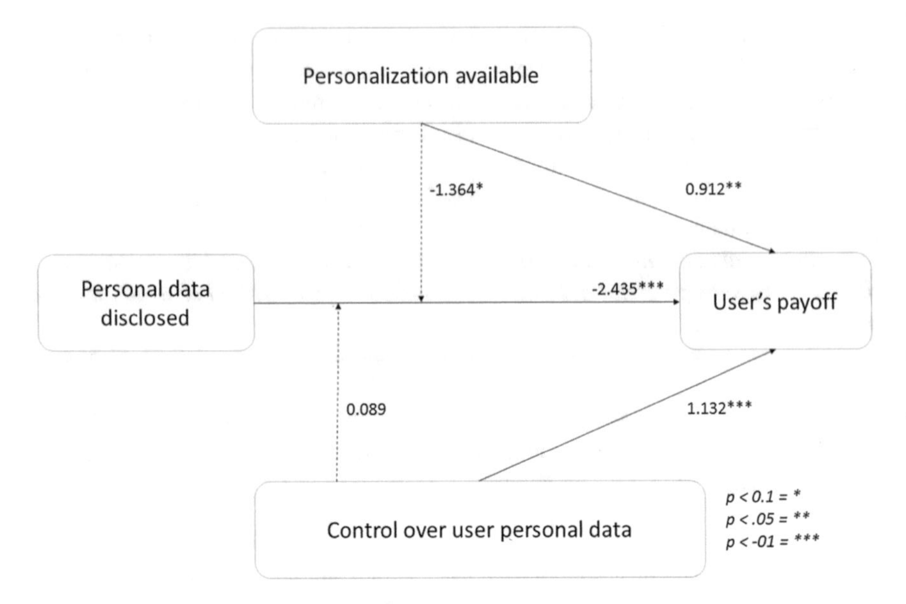

Fig. 2 The results of our survey study show the effect of personalization and control over the payoff of risk-neutral mobile users [26]

3.2 Business Model Insights

Figure 3 illustrates the results of our survey by means of a business model canvas. We have identified two types of users, associated to two types of strategies (data protection/no data protection). Accordingly, the firm can offer two types of value proposition (data protection/no data protection), even though, in the long term, if users and firms decide to protect the user's personal data, the overall system is more secure, since the personal data could be used as unique identifier. Accordingly, we conclude by saying that if a cooperation support system is in place, we could move from the notion of "security for privacy" towards the notion of "privacy for security" [7].

4 Step B: Travel to the Hotel by Car. How Tooxme Combined the Notions of Crowdsourcing and Dynamic Ridesharing

Let us imagine that Alice and Bob have to go from Geneva airport up to the village of Nax, in the middle of the Swiss Alps. They can choose among four possible options. They could do it by public transport, but they would have to switch trains and take a bus at the end, which would make the trip too long. They could rent a car, but they do

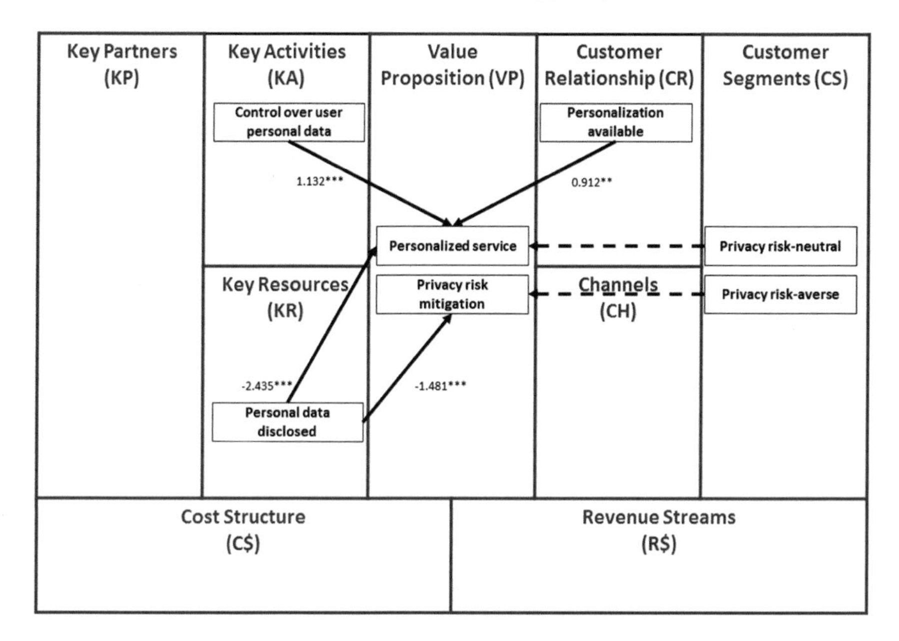

Fig. 3 Our results shown with a business model canvas [26]

not want to use a car during their stay. Finally, the taxi seems to be the most logical solution, but it might be too expensive. Therefore, the last option is to share a ride with someone else, which is going in the same direction. This solution has been put in place in many European cities, thanks to services that allow booking in advance a ride to share (ridesharing), such as Blablacar and Ouicar in France, or services that allows to book the ride few minutes before it takes place (dynamic ridesharing), such as Uber or Lyft in United States. The insights illustrated in this case have been obtained by working closely with a firm called Tooxme, in a project that took place between 2013 and 2015 in Western Switzerland to assess the feasibility of dynamic ridesharing. A famous example of dynamic ridesharing provider is Uber, and an example of business model canvas done for Uber can be easily found online (see for example, [29]). In this situation, drivers can decide to cooperate and share their ride and riders can decide to cooperate and accept the shared ride. Nonetheless, the perceived value of this system depends on how many riders and drivers are active at a given moment, which raises a question: how to obtain and maintain a large set of users on a cooperation support system?

4.1 Theoretical Model

The purpose of our artifact is to develop a process to monitor the diffusion of a dynamic ridesharing service by using a modified version of the model of susceptible-infected-recovered [13] illustrated in Fig. 4. Therefore, we are going to implement our model using software that is accessible to any company. In addition, we try to simplify our process as much as possible.

Our theoretical model is based on a simplified model which assumes that the recovered customers are susceptible to become infected again. Hence, we implement the SIS model, which uses four constructs: N, $n\%$, β and γ. The overall set of contacts is defined as N and the number of infected contacts is I_t. The number of susceptible contacts at any time t is $S_t = N - I_t$. The infected contacts represent the user base. Their initial amount is N*n% and their number changes over time according to the formula:

$$\begin{aligned} I_t &= I_{t-1} + new\ infected - new\ recovered \\ &= I_{t-1} + (\beta * I_t * S_t) - (\gamma * I_{t-1}) \\ &= I_{t-1} + [(\beta * I_t * (N - I_t)] - (\gamma * I_{t-1}) \end{aligned}$$

We collected our data from Google Trends (https://www.google.com/trends/). If we introduce the name of a company, Google trends returns the longitudinal representation of how many searches have been conducted for a particular term, relative to the total number of searches conducted on Google over time in a specific geographic area. We collected Google Trends data about Uber and Lyft from March 17, 2013 through May 4, 2014. The two CSV files were obtained by performing the two queries separately and by focusing on the United States. Nonetheless, we observed that if we run the query with the two keywords (Lyft and Uber), the ratio between the scores of the two companies on Google Trends in June 2013 (more or less 1:5) corresponds to the ratio between the amount of money the two companies obtained from investors.

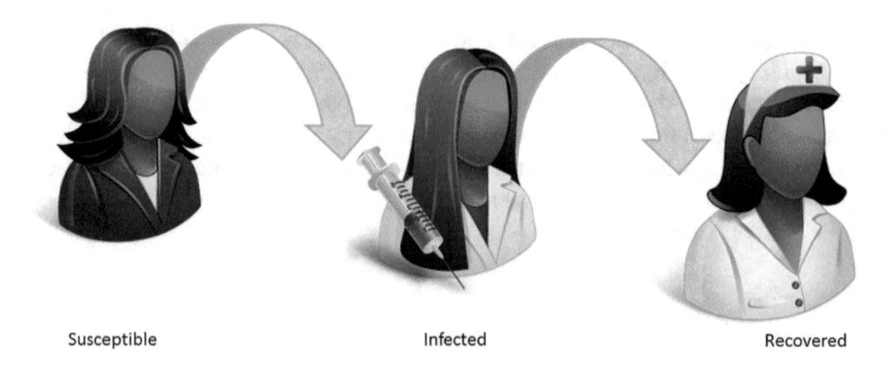

Fig. 4 Theoretical model of susceptible, infected and recovered model

This was assessed in August 2013, when Lyft funds reached USD 83 million against Uber's USD 410 million. This result might be coincidental, but if we assume that the final step of negotiation with venture capitalists often takes a couple of months, we can speculate that the Google Trends ratio between scores yields a proxy of the ratio between the user bases of the two firms. We also noticed the same results in France, when we compared the total funds obtained in December 2013 of 4.5 million Euros for Ouicar vs. the 12.5 million Euros for Blablacar (the ratio was 1:3). These results led us to our first testable proposition *(P1): The ratio of the Google Trends scores among firms is a proxy to assess their user base*. Once we optimized our beta and gamma constructs, the results we obtained confirmed our intuition. Our curves are able to explain the sudden increases in Google trends scores (acquisition of a new city), as well as the decrease (loss of user base in a city due to a high gamma value). Nonetheless, we wanted to verify if our beta and gamma were just random numbers or if they had some meaning. Indeed, one could assume that, if two firms with similar services operate in the same city, their gamma (which depends on the fit between the user base in that city and the firm) might be correlated, whereas the beta might not (since they depend on the company's customer relationship strategy). The correlation analysis between the beta and the gamma obtained shows that the two beta values are inversely correlated, whereas the gamma values are directly correlated. That leads us to our second testable proposition *(P2): The gamma values of competing dynamic ridesharing services in the same city are directly correlated, whereas the beta values are inversely correlated.*

4.2 Business Model Insights

Figure 5 shows how we could use β and γ to represent in the business model canvas the strategy to acquire and retain customers (the greater the value of beta, the more successful the campaign) and the fit between the value proposition of the firm and the type of customers in the city (the greater the gamma, the lower the fit). Those two elements are known to be crucial during the phases of customer discovery and customer validation, as defined by iterative approaches for business model definition [2].

Our results can be extended towards a fully decentralized solution as suggested by [35], who described a protocol for a fully decentralized P2P ridesharing management network that avoids centralized ride-matching agencies. They assessed its performance by using real mobility traces from the cabspotting project in the San Franscisco Bay area.

Accordingly, we described a possible extension of the business model for dynamic ridesharing [9], which involves the use of crowdfunding as a revenue model. This new service, called "Shared riders" as the inversion of "ride sharing", gathers every rider subscribed to the service to a network of casual drivers. This type of business would be initially funded by a crowd, which would be most likely the driver and the riders. For what concerns the business model sustainability, there is reason to believe that the service could be fairly efficient without needing a license, since there would be no

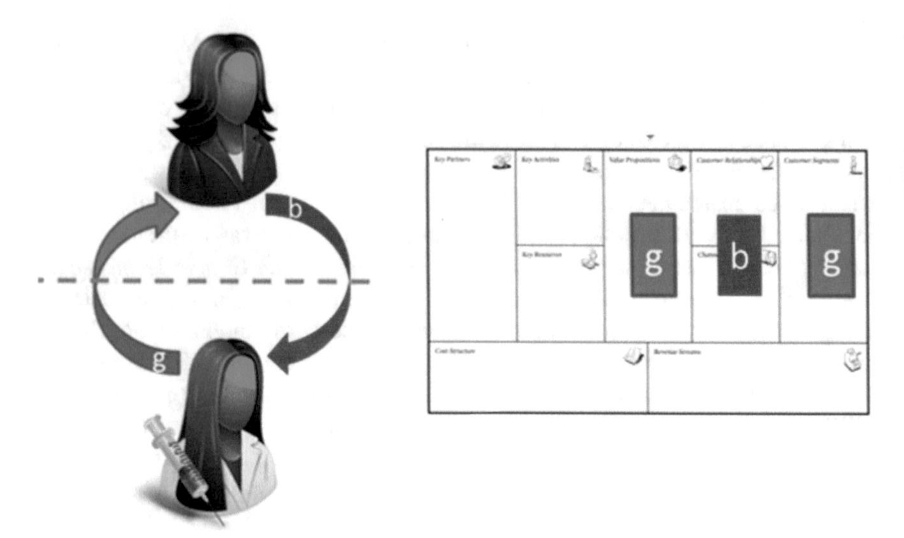

Fig. 5 Business model components that affect conversion rates among susceptible customers and infected customers

economic transaction among rider and driver. Indeed, the rider and driver would pay an initial fee to fund the service and then they would share rides and use the money in the shared fund to cover the cost. A tip can be expected as a form of compensation to exceptional drivers, but that it is not considered an economic transaction.

5 Step C: Hotel Check-In. Innovation by Design in Hospitality Management at Maya Boutique Hotel

Let us imagine that Alice and Bob have arrived to the Maya Boutique Hotel in Nax, a luxury hotel next to Sion (more information on maya-boutique-hotel.ch). The Maya Boutique Hotel is currently redefining its strategy for acquiring new customers after having won the *Best Innovation in Hotel Concept* prize by the [24] and the *Sustainability* prize by the [12]. To help the owners of the hotel, we used design thinking tools to find insights for new value propositions and new products to develop. We did so by formalizing the assumptions made about the hotel guests, by analyzing the spatial distribution of customers, by doing quantitative tests on client data and company data from the site and by completing quantitative tests with qualitative interviews to understand more deeply a customer segment. In this case, a service provider (for example, the hotel owner) can decide to cooperate with another service provider (for example, a restaurant owner or the hotel owner of another location). A famous example of sharing economy for hospitality management is Airbnb, and examples of business model canvas done for Airbnb can be easily found online, for example [30]. Nonetheless, there are few examples of hotels looking for alternative solutions to reduce their dependencies from Online Travel Agencies (OTA), such as Booking.com, and to move towards a partially decentralized solution. Previous studies have already described a decentralized and co-utile reputation system, which would allow circumventing centralized platforms [19]. Hence, we shall focus on the business model of this type of service.

5.1 Theoretical Model

Figure 6 illustrates a process that allows hotels to set up a strategy of acquisition and additional loyalty, partially freed from OTAs. By observing the online behavior of its client, the hotel can improve its strategy for customer acquisition. The hotelier will be able to use different alternatives to online booking as Triptease, Seedka or the new service Bookbedder.com. Then, if the client has enjoyed the hotel, the facilities and location, the hotel owner should include him in a loyalty program. If the customer has enjoyed only the hotel and activities, it is preferable to use a collaborative CRM to propose another location corresponding to expectations of the customer. If the customer has enjoyed the hotel but looks for different activities, the hotelier can choose a tool to dynamically bundle activities into a set of modular offers.

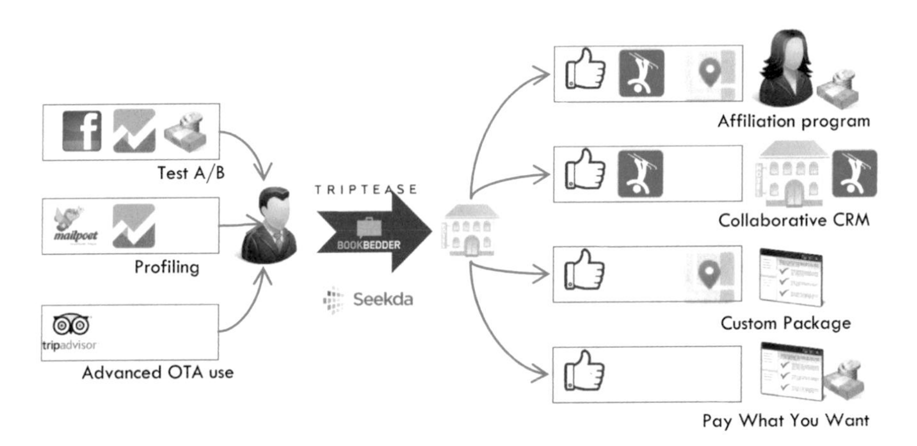

Fig. 6 Hotel discovery process, reservation and affiliate programs - theoretical model and results of a qualitative study

A team of the Service Design Lab of HES-SO conducted a dozen interviews with hotel guests and potential customers who reside around the Geneva Lake. Generally speaking, respondents perform an extensive research on the internet before booking their room for a romantic holiday. The choice of hotel is defined primarily by the location, time and dates. Pictures, descriptions, rating and especially the comments of former clients become formidable tools in decision making. Results found on Booking.com, combined with analysis of information on TripAdvisor has become an absolute must in order to plan and choose the hotel offering the best services, benefits, price and experience. For a romantic stay, respondents look for a place with beautiful surroundings and breathtaking views. The hotel must have a particular style with spacious rooms with all amenities and services expected of an establishment of high end. They prefer hotels with a spa and wellness. The food is also important: something in the discovery and surprising, out of the ordinary. Respondents already spent romantic weekends in hotels and, for the most part have had good experiences. But they feel that packages do not fully meet their needs, and that such bundled offers should be customized by the client. In this sense, a flexible or adaptable package would be more interesting. Respondents know that customer acquisition and retention are key priorities for the hotel and they gladly make recommendations if the experiment is successful. But they feel overwhelmed by the number of loyalty programs on the market. In an upscale hotel, respondents prefer a softer and less direct approach. To attract new customers to join the loyalty program it is sometimes better to offer something of value to them as a free service, a meal or a bottle of champagne, instead of a prize reduction.

5.2 Business Model Insights

Figure 7 illustrates how some clients of the cooperation support system can be at the same time key partners, since they use and offer services at the same time.

Moreover, this type of strategy leads to the creation of clusters within the overall network of service providers. Indeed, some service providers will decide to cooperate among them, but not to cooperate with others. Accordingly, the analysis of the overall network might allow the cooperation support system to enhance cooperation among agents by adding and removing links among agents. This strategy has been explained in a research paper [5], but it has not been empirically tested. Thus, it is beyond the scope of this study.

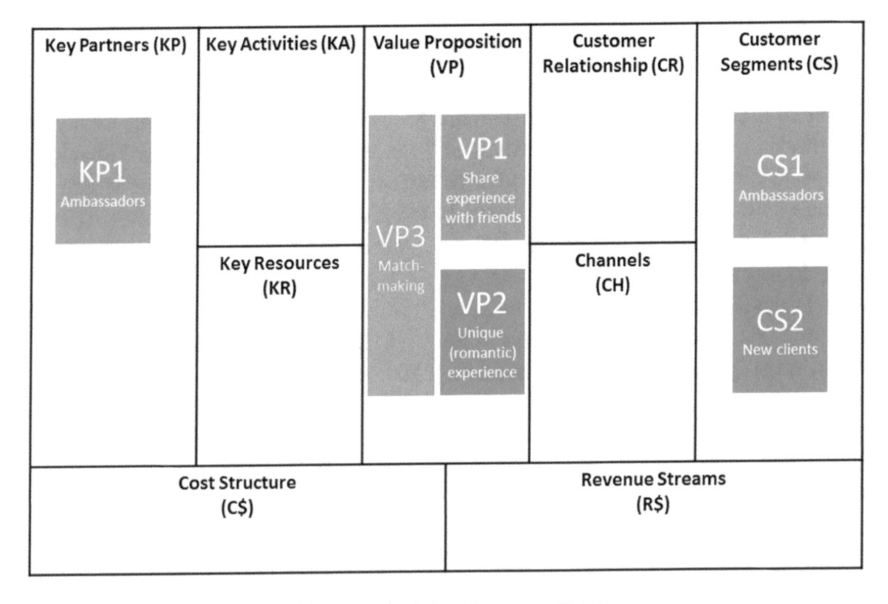

Fig. 7 Simplified business model canvas for Maya Boutique Hotel

6 Step D: Dinner Out. Healthy Lottery as an Example of Gamification to Increase Patient's Adherence to Medications

Let us imagine that Alice and Bob decide to go out for dinner. Bob has been recently diagnosed type-II diabetes, which requires him to increase the amount of physical exercise done every day while trying to follow the diet prescribed by his doctor. The compliance rate of long-term medication therapies can be estimated between 40 and 50%. The rate of compliance for short-term therapy is much higher (between 70 and 80%), while the compliance with lifestyle changes is the lowest (20–30%) [18]. The analysis of four systematic reviews of reward-based financing has found that financial incentives targeting recipients of healthcare is effective in the short run for simple and distinct, well-defined behavioral goals, whereas there is less evidence that financial incentives can sustain long-term changes [32]. Since people place more weight on the present than the future costs, a system designed to increase the immediate rewards may influence people's propensity to act, by lowering the social and economic factors that lead to non-compliance. Hence, studies on behavioral economics emphasize (a) the importance of frequent feedback and incentives, (b) the motivational power of lotteries regarding other financial features and (c) the motivation force of anticipated regret [38].

In this situation, the patient can decide to cooperate with the doctor and comply with prescription, whereas the doctor can decide to cooperate and adapt prescription to user's specific needs to increase his motivation.

6.1 Theoretical Model and Results of a Simulation

Financial incentives for diabetes self-management have only begun to be explored. Individuals expect financial incentives to be a stronger motivation for behavioral change [3], even though incentives for behaviors are preferred for tasks which are considered less challenging [4]. Lottery-based incentives improve monitoring rates among patients with uncontrolled diabetes, and it seems that the smaller expected value lottery ($1.40 per day) is considerably more effective in the post-incentives period than the larger expected value lottery ($2.80 per day) [37]. Our design theory has four constructs to describe the system: (1) the short-term and the long-term evolution of the patient's clinical situation, (2) the monetary incentives, (3) the change in the healthcare provider efficiency and (4) the change in the motivation of the patient. The short-term improvement of the patient's condition can be measured by the adherence to medication and the meetings with the healthcare provider, which are reported in the patient's logbook. The sustainable change of the patient can be measured by mobile applications that monitor (a) the level of blood glucose, (b) the level of HbA1c and (c) the Body Mass Index. The monetary incentives are measured by the money delivered to the patient. Finally, the healthcare provider efficiency can be measured by the average amount of hours spent with the patient, whereas the patient's motivation can be measured by a survey that assesses intrinsic and extrinsic motivation [34].

Figure 8 represents the two functions of the system by using two rectangles. The healthcare provider and the patient affected by diabetes meet to set the goals, in terms

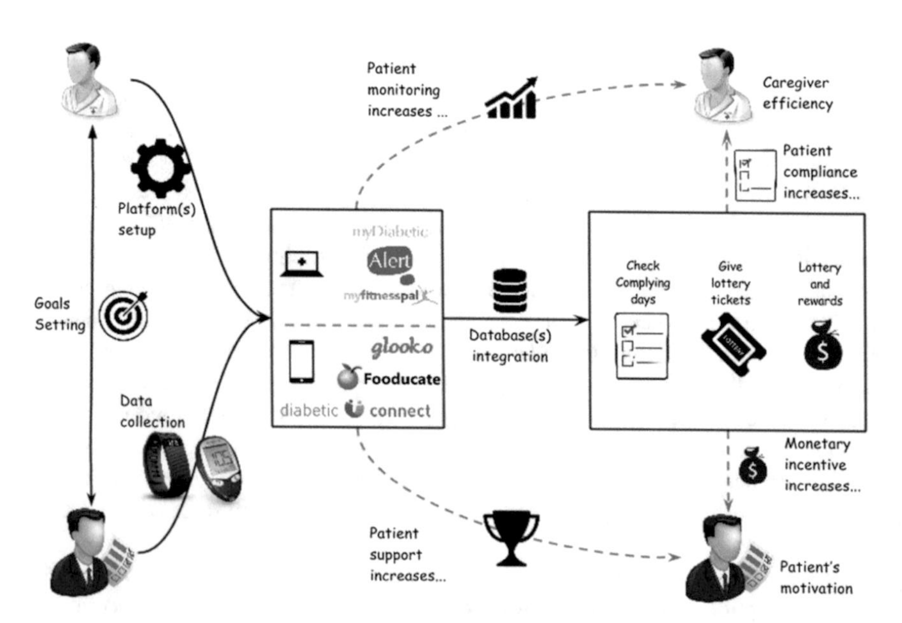

Fig. 8 Description of how goals are set, data is analyzed, and rewards are given (adapted from Bonazzi et al. [6])

of diet, exercise, medications and smoking cessation. We assume that the patient uses a set of devices to automatically collect data every day, whereas we also expect the healthcare provider to spend some time to set up the platform at the beginning. By using our four constructs, we derive two sets of propositions. As a starting point, we need to prove that the platform is economically viable without being profitable, due to ethical concerns and legal requirements in some countries. As it turns out, most lottery systems were designed to be fairly profitable and their approach needed to be modified to fulfill our expectations. Therefore, our platform is conceived to assure that compliant participants will receive the service for free, while collecting enough resources to cover operational costs.

P1: *Monetary incentives* will cover the *inscription cost* of participants that comply with doctor prescriptions.
We also believe that the random rewards delivered by e-mail will motivate the patients without requiring any additional effort on their side. Moreover, we seek to increase patients' self-management on the long term.

P2: The way *monetary incentives* are used in the system will increase *the patient's intrinsic motivation.*

To test our first proposition, we ran a simulation with random numbers for 10,000 participants over 52 weeks. Figure 9 illustrates the distribution of the yearly revenues. The median of our simulated results is close to zero (a participant can expect to receive back the money initially spent, which is \$1 per week or \$52 in one year). Moreover, everyone is expected to win something, which we call *YR* for yearly revenues and that should be between \$22 (=\$52 − \$30) and \$82 (=\$52 + \$30). Therefore, proposition P1 (*monetary incentives will cover the inscription cost of participants that comply with doctor prescriptions*) appears to be validated.

Nonetheless, in our subsequent simulations we took into account that the compliance degree among patients with type II diabetes cannot be expected to be normally distributed, due to the effort required to change the lifestyle of a person. Therefore, we created a set of matrices PART with random numbers generated by using a Poisson distribution, whose formula is

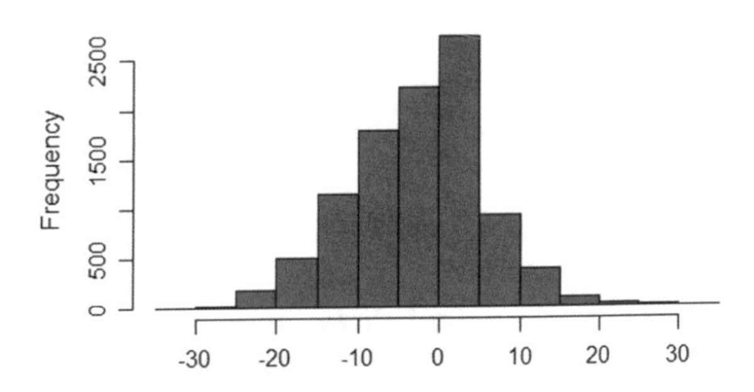

Fig. 9 Distribution of yearly revenues *YR* (USD VS frequency)

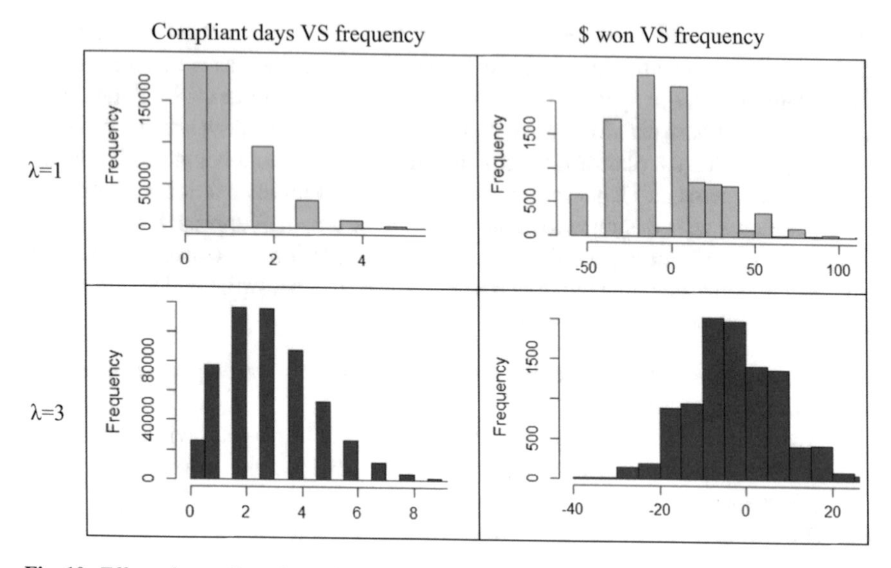

Fig. 10 Effect of compliant days distribution over money won distribution

$$P\text{ (k events in the interval)} = \frac{\lambda^k e^{-\lambda}}{k!}$$

Then, we observed how the results changed as we modified the λ. Our simulation shows what would happen with a population of patients that comply less than 2 days per week ($\lambda = 1$), which is compared to a population that complies up to 7 days per week (when a simulated participant complies more than 7 days per week, the compliance degree is set at 7 days over 7 days).

Further analyses illustrate that, when most people do not comply ($\lambda = 1$), those who comply can expect to win a significant amount of money at the end of the year. Nonetheless, as a growing number of participants become motivated to comply ($\lambda = 3$), the weekly rewards get smaller and the expected YR gets closer to the one shown in Fig. 10. Therefore, the system is meant to use extrinsic motivation as an enabler for intrinsic motivation. Therefore, proposition P2 (*the way monetary incentives are used in the system will increase the patient's intrinsic motivation*) appears to be validated.

6.2 Business Model Insights

Figure 11 illustrates the business model canvas associated to the fourth case, which is meant to possess the overall features of the other three cases. Here is the detailed description of each of the nine elements.

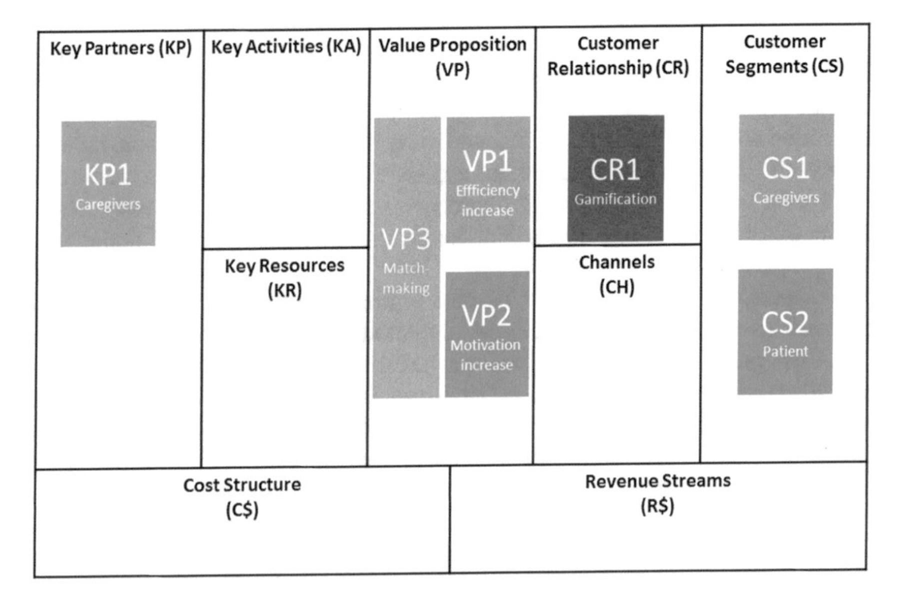

Fig. 11 Business model of our mobile service

Customer segments. For sake of simplicity, we focus here on two main customer segments: (a) patients and (b) caregivers/healthcare providers. Patients should be considered both as users and clients. Indeed, they should invest some money in order to be motivated to improve their lifestyle. In our simulation, we have suggested that a community of 10,000 patients can be reached, even though that will require a significant amount of effort to promote the service and to manage the community of patients. In comparison, a service such as "Patient like me" claims to have 380,000 patients providing health data on over 2,500 diseases. Therefore, healthcare providers would be key players for the success of the service, by promoting the mobile service among their patients in order to reduce the number of hospitalizations due to lack of compliance.

Healthcare providers can choose to receive a report about patients' compliance, as a complement to the information already offered by other applications such as myDiabeticAlert. Finally, an insurance firm could decide to sponsor this service as a way to reduce the cost of hospitalization, in their own interest and in the interest of their clients. Thus, the proposed lottery incentives yield co-utility between patients, healthcare providers and insure companies. Nonetheless, in the following paragraphs we assume that the insurance does not pay for this service.

Value propositions. The service offers a new way to use extrinsic motivation to improve the patients' compliance and to increase the doctor's efficiency. This idea is presented to patients as a tool that properly rewards their efforts, whereas it is presented to healthcare providers and insurance firm as a way to reduce the costs of multiple hospitalizations.

Customer relationship. The service offers feedback every week to every patient by delivering a reward, even though the interaction with patients is fairly standardized to reduce operational costs and to focus on the "job to be done". On the one hand, the interaction with patients and healthcare providers is kept to the minimum in order to not give the impression of an additional application (the overall systems needs to be initially set up and then it automatically collects data and gives updates every week). On the other hand, the interaction with the insurance firm is extremely limited due to privacy concerns.

Channels. Patients are reached by online promotion or advised by healthcare providers (for the sake of simplicity, we do not take into account patients associations), whereas professional caregivers can be reached with specialized publications and talks at professional gatherings. The service is delivered to patients and healthcare providers by e-mail, as a complement to dashboards already offered by other mobile applications. Insurance firms do not really interact with the system and are most likely interested in presentations about the overall performance of the system.

Key activities. A crucial task for the success of the platform consists in the acquisition and retention of patients, since the simulation has shown how the average cost for each participant gets closer to zero as soon as the degree of compliance among participants resembles a normal distribution. Hence, marketing efforts and community management have the highest priority as well as the tasks required to assure that the system can automatically collect and analyze the sensor data about each patient.

Key resources. On the one hand, the lottery system and the know-how associated to its way of working are a valuable set of resources. On the other hand, the community of patients and healthcare providers is what keeps the overall service alive. Moreover, we list the patient's data among the key resources, as a reminder of all the privacy concerns associated with medical records. Patients' data are initially collected and stored by other applications and one could decide to erase the data every week, once the winners are declared, as long as there is no need to audit the lottery system afterwards.

Key partners. The patients and healthcare providers are important partners to promote the service among patients; hence, they should be considered both partners and customers. Moreover, professional caregivers could take part as medical advisors during the research and development process. Smartphone applications that collect data about the patients are key partners, since they allow reducing the cost for data collection and software development. Nonetheless, the risk associated to outsourcing these tasks is related to the high degree of uncertainty concerning the evolution of the ecosystem of these applications, both in terms of merge and acquisition and in terms of survival rate. This could lead to significant cost for constant adaptation of the interfaces between our platform and the other applications.

Cost structure. A large part of the inscription fees goes back to the patients. The remaining part of the revenues should be used for marketing and sales and for research and development. Our cost structure assumes a small team of five people, which has externalized most of the tasks.

Revenue model. Assuming that the service reaches a community of 10,000 participants, one could expect the patients' inscription fee to shift from \$1 per week to

$1 per day, leading to a revenue flow of some $180,000 from the 5% commission rate. Indeed, even though a subscription fee of $365 per year might discourage a larger amount of patients, this cost is less than the subscription for a gym and our simulation shows that compliant patients will have most of their money back at the end of the year. If we assume an inscription fee of $1 per day, it would be possible to devolve a certain amount of collected money to support social projects, in accordance to Swiss regulations. The revenue flow associated to the doctors might seem reasonable but it will not bring a significant amount of revenues until the system has been tested according to hospital standards. In the first year, some money could come from research projects to assess the feasibility of the model, but this should not be seen as a sustainable solution. Therefore, a partnership with some insurance firms might be required, even though that option brings questions related to privacy concerns and potential conflicts of interest.

7 Future Directions of Investigation

In this chapter, we have used the business model canvas to represent a set of guidelines to design the business model for a cooperation support system. Four cases have been used to derive strategic insights from work experience.

We conclude our chapter by listing three directions of investigation related to our cases, which are expressed under the form of statements for interesting research [17]:

1. *Causation*: What seems to be the independent phenomenon in a causal relation is in reality the dependent phenomenon. The first case has described how we could use a cooperation support system to security as an emerging phenomenon, moving from "security for privacy" towards "privacy for security". How many other examples of emerging behavior can be associated to co-utile protocols?
2. *Stability*: What seems to be a stable and unchanging phenomenon is in reality an unstable and changing phenomenon. The second case has described how agents might change their strategy depending on the number of users of the cooperation support system. How to formalize a co-utile protocol in this type of situations?
3. *Generalization*: What seems to be a general phenomenon is in reality a local phenomenon. The third case has described how agents might create clusters within the network and decide to use a co-utile protocol only with some players. How can the cooperation support system affect the topology of the network in order to enhance cooperation?

While this research focuses on the number of users involved and the decentralization of the information system, we conclude this section by presenting two broader investigation paths beyond our cases.

On the one hand, the analysis of other elements of the business model, such as key resources and key activities, mobilized in the co-utility protocol, could be part of the definition of the financing mode of the systems. This leads us to the following question: How are key resources and key activities involved in the co-utility process?

On the other hand, according to the effective reasoning approach [36], key activities could be the starting point for identifying opportunities. While effective reasoning begins with a given set of means allowing goals contingently, it is complementary to the causal reasoning, which begins with a specific goal before identifying the best way to achieve it. So far, a few local examples in Western Switzerland develop co-utile solutions, based on key latent resources and key activities, and answer to the effectual reasoning principles:

1. The partnership between three hotels and a camping in the creation of unique marketing campaigns: the effectual reasoning approach depends upon strategic alliances, instead of competitive analyses.
2. The orchestration of jam production combining saving of discontinued raw materials, the environmental education of children and the participation of local cooks: the effectual reasoning approach emphasizes affordable loss, instead of expected return.
3. The local grocery store chain benefiting from the closure of the post offices in the villages to extend its local services: the effectual reasoning approach stresses the leverage of contingencies, instead of exploiting the pre-existing knowledge and prediction.

According to [17], this research interest is related to the characterization of a single phenomenon related to the organization: what seems to be a disorganized (unstructured) phenomenon is in reality an organized (structured) phenomenon. This leads us to the following question: How can key resources and key activities lead companies to the identification of cooperation opportunities?

References

1. Anderson, R.: Why information security is hard-an economic perspective. In: Computer Security Applications Conference (2001)
2. Blank, S.: The startup owner's manual: the step-by-step guide for building a great company. BookBaby (2012)
3. Blondon, K., Klasnja, P., Coleman, K., Pratt, W.: An exploration of attitudes toward the use of patient incentives to support diabetes self-management. Psychol. Health **29**, 552–563 (2014). doi:10.1080/08870446.2013.867346
4. Blondon, K.S.: Patient attitudes about financial incentives for diabetes self-management: a survey. World J. Diabetes **6**, 752–758 (2015). doi:10.4239/wjd.v6.i5.752
5. Bonazzi, R., Buesser, P., Holzer, A.: Cooperation support systems for open innovation. In: Proceedings of WOA (2012)
6. Bonazzi, R., Cimmino, F.M., Blondon, K.: Healthy lottery. An economically viable mobile system to increase compliance of individuals with diabetes. In: Proceedings of the 50th Hawaii International Conference on System Sciences. Kona, HI (2017)
7. Bonazzi, R., Fritscher, B., Liu, Z., Pigneur, Y.: From "security for privacy" to "privacy for security." In: 2011 15th International Conference on Intelligence in Next Generation Networks (ICIN), pp. 319–324. IEEE, New York (2011)
8. Bonazzi, R., Pigneur, Y.: The Hitchhiker's guide to the galaxy of dynamic ridesharing. In: 2015 48th Hawaii International Conference on System Sciences (HICSS), pp. 1207–1216. IEEE, New York (2015)

9. Bonazzi, R., Poli, M.: Beyond Uber. Business model considerations for alternatives to traditional taxis. In: Proceedings of Itais (2014)
10. Bonazzi, R., Schegg, R.: An alternative to online travel agencies for retention of hotel customers. In: Proceedings of Entrenova 2016. Rovinj, Croatia (2016)
11. Bonneau, J., Preibusch, S.: The privacy jungle: On the market for data protection in social networks. In: Economics of Information Security and Privacy, pp. 121–167. Springer, Berlin (2010)
12. Boutique Hotel Awards. Sustainability - Boutique Hotel Awards. In: BoutiqueHotelAwards.com (2015). http://www.boutiquehotelawards.com/hotel-experiences/sustainability-2015/. Accessed 1 Mar 2016
13. Cannarella, J., Spechler, J.A.: Epidemiological Modeling of Online Social Network Dynamics (2014)
14. Chellappa, R.K., Sin, R.G.: Personalization versus privacy: an empirical examination of the online consumer's dilemma. Inf. Technol. Manag. **6**, 181–202 (2005)
15. Danezis, G., Domingo-Ferrer, J., Hansen, M., et al.: Privacy and Data Protection by Design-from Policy to Engineering (2015)
16. Das, T.K., Teng, B.-S.: Trust, control, and risk in strategic alliances: an integrated framework. Organ. Stud. **22**, 251–283 (2001)
17. Davis, M.S.: That's interesting: towards a phenomenology of sociology and a sociology of phenomenology. Philos. Soc. Sci. **1**, 309 (1971)
18. DiMatteo, M.R.: Patient adherence to pharmacotherapy: the importance of effective communication. Formul Clevel Ohio **30**, 596–8 (1995)
19. Domingo-Ferrer, J., Farràs, O., Martínez, S., et al.: Self-enforcing protocols via co-utile reputation management. Inf. Sci. **367**, 159–175 (2016)
20. Domingo-Ferrer, J., Martínez, S., Sánchez, D., Soria-Comas, J.: Co-utility: self-enforcing protocols for the mutual benefit of participants. Eng. Appl. Artif. Intell. **59**, 148–158 (2017)
21. Gregor, S.: The nature of theory in information systems. MIS Q. 611–642 (2006)
22. Herrmann, D.S.: Complete Guide to Security and Privacy Metrics: Measuring Regulatory Compliance, Operational Resilience, and ROI. CRC Press, Boca Raton (2007)
23. Hevner, A.R., March, S.T., Park, J., Ram, S.: Design science in information systems research. MIS Q. **28**, 75–105 (2004)
24. Hospitality Awards.: Best innovation in hotel concept. In: HospitalityAwards.com (2015). http://hospitalityawards.com/en/the-categories/hospitality-awards/2015/best-innovation-in-hotel-concept/. Accessed 1 Mar 2016
25. Lee, A.: Researchable directions for ERP and other new information technologies. MIS Q. **24**:III (2000)
26. Liu, Z., Bonazzi, R., Fritscher, B., Pigneur, Y.: Privacy-friendly business models for location-based mobile services. J. Theor. Appl. Electron. Commer. Res. **6**, 90–107 (2011)
27. Miltgen, C.L.: Disclosure of personal data and expected counterparties in e-commerce: a typological and intercultural approach. Syst. Inf. Manag. Fr. J. Manag. Inf. Syst. **15**, 45–91 (2010)
28. Nalebuff, B.J., Brandenburger, A.M.: Co-opetition: competitive and cooperative business strategies for the digital economy. Strategy Leadersh. **25**, 28–33 (1997)
29. Nextjuggernaut.com. Uber Business Model Canvas: know what led to Uber's success. In: Nextjuggernaut.com (2015). http://nextjuggernaut.com/blog/uber-business-model-canvas-what-led-to-uber-success/. Accessed 15 Dec 2017
30. Nextjuggernaut.com. How Airbnb works insights into business & revenue model. In: Nextjuggernaut.com (2015). http://nextjuggernaut.com/blog/airbnb-business-model-canvas-how-airbnb-works-revenue-insights/. Accessed 15 Dec 2017
31. Osterwalder, A., Pigneur, Y.: Business Model Generation: A Handbook for Visionaries, Game Changers, and Challengers. Wiley, London (2010)
32. Oxman, A.D., Fretheim, A.: An Overview of Research on the Effects of Results-Based Financing (2008)
33. Palen, L., Dourish, P.: Unpacking privacy for a networked world. In: Proceedings of the SIGCHI Conference on Human Factors in Computing Systems, pp. 129–136. ACM, New York (2003)

34. Pelletier, L.G., Tuson, K.M., Haddad, N.K.: Client motivation for therapy scale: a measure of intrinsic motivation, extrinsic motivation, and amotivation for therapy. J. Pers. Assess. **68**, 414–435 (1997)
35. Sánchez, D., Martínez, S., Domingo-Ferrer, J.: Co-utile P2P ridesharing via decentralization and reputation management. Transp. Res. Part C Emerg. Technol. **73**, 147–166 (2016)
36. Sarasvathy, S.D.: Causation and effectuation: toward a theoretical shift from economic inevitability to entrepreneurial contingency. Acad. Manag. Rev. **26**, 243–263 (2001)
37. Sen, A.P., Sewell, T.B., Riley, E.B., et al.: Financial incentives for home-based health monitoring: a randomized controlled trial. J. Gen. Intern. Med. **29**, 770–777 (2014). doi:10.1007/s11606-014-2778-0
38. Volpp, K.G., Loewenstein, G., Troxel, A.B., et al.: A test of financial incentives to improve warfarin adherence. BMC Health Serv. Res. **8**, 272 (2008). doi:10.1186/1472-6963-8-272

The Need of Co-utility for Successful Crowdsourcing

Enrique Estellés-Arolas

Abstract Technological development has promoted the rise and use of the collective intelligence through Internet. To efficiently handle this collective intelligence, several processes have naturally emerged. Crowdsourcing is one of them. By using crowdsourcing, the undertaking of a task can be proposed by a person or organization to the crowd that composes Internet. Although these proposed tasks could vary in their requirements for successful accomplishment, any crowdsourcing initiative always includes different benefits for the promoters of the initiatives and one or more rewards for each person of the crowd. These rewards play a key role because their evaluation by the crowd will condition the number, the interest and the implication of participants in the initiative. To design and model the interaction between the crowd and the initiative promoter, game theory appears as a suitable tool. In this chapter, the close relationship between crowdsourcing and co-utility, a type of collaborative interaction defined in terms of game theory, will be studied. It will be shown that one of the easiest ways to achieve successful crowdsourcing initiatives is to give them a co-utile configuration.

1 Introduction

Collective intelligence has been a fact since human beings relate to and cooperate with each other. This cooperation has always had a specific goal, a purpose, which is what allows us to categorize specific behaviours as 'intelligent' [33].

In the historical development of the collective intelligence, the emergence of the Internet represented a fundamental change. Whereas before the Internet collective intelligence met serious limitations to gather a great amount of people –organization, space, management restrictions, etc., the emergence and growth of the Internet, and the virtual space it implies, made it posible to overcome many of those restrictions. Obviously, this change also originated new challenges and problems.

E. Estellés-Arolas (✉)
Catholic University of Valencia San Vicente Mártir, C. Quevedo 2, 46001 Valencia, Spain
e-mail: enrique.estelles@ucv.es

J. Domingo-Ferrer and D. Sánchez (eds.), *Co-utility*, Studies in Systems, Decision and Control 110, DOI 10.1007/978-3-319-60234-9_11

Even though it is true that the Internet allowed a great amount of people to connect technologically, the way of doing so efficiently was still to be found. Given the vast potential of collective intelligence, it was needed to design interactions between the people and the technology that made their connections possible in the most cost-effective way. This need motivated the human computer interaction discipline, whose purpose is to design and generate digital channels that make the emergence of the collective intelligence possible.

These channels were materialized in what is known as web 2.0. This new paradigm, which arose at the beginning of this century, reshaped the very collectivity it connected, by turning the spectator into a main actor. In fact, from the lists of platforms categorized by O'Reilly [28] in the workshop where the term "web 2.0" was coined, most of them were plataforms related to collective intelligence - Wikipedia, Flickr, tagging platforms, etc. The common point, as O'Reilly would say, is that in all those platforms the users were given the opportunity to create, contribute and generate. That is, the possibility to participate.

This collective intelligence that used platforms of the web 2.0 as a means started to generate different initiatives that required a kind of process or methodology in order to manage them effectively. This is how the so-called 'open innovation' appeared. This term, coined by Chesbrough [6], refers to the opening of companies to valuable ideas that come from the outside, and to valuable ideas from the inside of the company reaching the outer market through other companies (that obtain a benefit in return, obviously). Another methodology is co-creation, which has been defined by Banks and Potts [1] as an agreement between a promoter of a project and a collectivity of users that want to contribute significantly to that project, in an environment based on transparency and mutual acknowledgment.

There are many other methodologies, but the one we are interested in in this chapter is crowdsourcing, which has become more and more important. Crowdsourcing is a term coined by the U.S. journalist J. Howe in [17]. It refers to the undertaking by a crowd of a task proposed by a promoter, in exchange for a reward.

In the three above-mentioned examples (open innovation, co-creation, crowdsourcing), there are two common characteristics – the opening to the outside and the acknowledgment for the work done. In the end, it is all about the interaction between two or more entities (a promoter and an undetermined group of people) in exchange for a reward.

Within this framework, game theory turns out to be a useful tool to express, study and design the interaction between the promoter and the group of people.

Game theory is the mathematical study of interaction among independent, rational and self-interested agents. It is mainly studied by mathematicians and economists, being the micro-economy its main field of application. Anyway, due to its object of sudy, nowadays it is being applied in such diverse fields as the social sciences [6], biology [10], humanitarian aid [25] or computer networks [24].

In the case of collective intelligence, the promoter as well as the crowd are rational and self-interested agents. That is, they are agents that, when facing a specific situation, have some preferences and act accordingly to achieve their goals following a determined strategy. The agents' interests can be modelled and quantified through utility theory, which quantifies an agent's degree of preference across a set of available alternatives [22].

The interaction between the agents is defined according to protocols that specify the expected behaviour from both the agents and the communication patterns available. Obviously, in many cases, the interests of the agents involved will not agree, so the design of those interactions or protocols will not be a simple task. That is the reason why good designs of protocols that are followed by the agents voluntarily are essential: these are called self-enforcing protocols. More specifically, a protocol is self-enforcing if no agent has any rational incentive to deviate from it provided that the other agents stick to the protocol.

According to Domingo-Ferrer et al. [8], a self-enforcing protocol is co-utile if the way for a participating agent to increase her own utility is to help other protocol participants to increase their utilities.

This chapter will focus on the extent to which co-utility is applicable to the interactions that arise in collective intelligence initiatives, specifically, in crowdsourcing. We first give background on crowdsourcing and game theory. Then the way co-utility could be applied in a general approach to crowdsourcing initiatives will be analyzed.

2 Crowdsourcing: The Concept

From the moment it was 'born', crowdsourcing has followed a logical evolution. At first, Jeff Howe defined it as a step forward from the classic outsourcing, a definition accepted and slighlty qualified by Brabham [3]. According to Howe [17], crowdsourcing is "*the act of a company or institution taking a function once performed by employees and outsourcing it to an undefined (and generally large) network of people in the form of an open call. This can take the form of peer-production (when the job is performed collaboratively), but is also often undertaken by a sole individual*".

In Estellés-Arolas and González-Ladrón-de-Guevara [11], the authors took another step forward by suggesting a definition that incorporated the one suggested by Howe [17] and also amplified and qualified it. It is worth pointing out that in this definition there are eight components that make the distinction between the crowdsourcing and other types of initiatives possible. The authors claim that in any crowdsourcing inititive the following must be present:

1. A crowdsourcer that promotes the initiative;
2. A specific task to be done with a clear objective;
3. A crowd that will carry out that task;
4. A reward to be given to the crowd in exchange for that task;
5. A benefit to be perceived by the crowdsourcer for the execution of that task;
6. A participative process that requires a conscious activity by the crowd;
7. An open call, so that anyone can participate,
8. Internet as a means of interaction.

Apart from these eight elements, these authors draw the conclusion of reciprocity, i.e., for a crowdsourcing initiative to be developed, there needs to be a mutual benefit between the crowdsourcer and the crowdworkers, a win-win relationsip.

Some authors suggest what could be considered a step forward in this direction. Such is the case of de Bigham et al. [2]. These authors suggest the existence of three different types of crowdsourcing:

(1) Directed crowdsourcing. It is the classic crowdsourcing, where a crowdsourcer gathers some people to carry out a task. Amazon Mechanical Turk would be a good example of the platforms applying this type of crowdsourcing.
(2) Collaborative crowdsourcing. In this case, it is the crowd itself that determines its own organization and work. The members of the crowd tend to share the same interests and this type of initiatives emerge, to a certain extent, spontaneusly.
(3) Passive crowdsourcing. The result obtained in this type of initiatives is part of the regular behaviour of the crowd. An example of this type of crowdsourcing would be the one suggested by Sambuli et al. [29], where messages on Twitter to predict political outcomes were followed and analyzed.

The new proposal by Bigham et al. [2] would be based on the blurring of the figure of the crowdsourcer (being identified, on some occasions, with the crowd itself, as in the case of collaborative crowdsourcing) and the elimination of the requirement of a conscious action by the crowd (as in the case of passive crowdsourcing, where the crowd does *not* actually perform a task in response to an open call).

Although the first contribution of these authors (i.e., the consideration of the crowd playing the role of the crowdsourcer) can be discussed but accepted as a step forward, the second contribution would require a deeper debate which would exceed the limits of this chapter. Even though it is a clear case of collective intelligence, it would be necessary to analyze whether it must be categorized as crowdsourcing or not.

Regardless of potential disagreements, in those three contributions there are some clear elements that appear in any type of crowdsourcing initiative and allow delimiting the concept – a series of participants (the crowd and the crowdsourcer), a task and a reward.

2.1 *Different Ways to Use Crowdsourcing*

Since the term crowdsourcing was coined by Howe in [17], different typologies of crowdsourcing initiatives have been suggested. Some of them are centered on a specific area, such as the contribution by Oomen and Aroyo [27] for cultural heritage, whereas others are more general, such as the ones proposed by Howe [18], Brabham [3] or Geiger et al. [13]. The comprehensive proposal by Estellés-Arolas et al. [12] can be included in this last group; it distinguishes five crowdsourcing initiatives, depending on the task to be done:

1. Crowdcasting. This is a contest-like crowdsourcing initiative, in which only one (the best or the first) of the members of the crowd will receive the reward.
2. Crowdcollaboration. This is a crowdsourcing initiative in which communication between crowd members occurs to support each other or for particular online brainstorming initiatives.
3. Crowdcontent. In an initiative of this type, the crowd uses its labour and knowledge to create, analyse or search for content of different types in different media (documents, images, the Internet, etc.)
4. Crowdfunding. In this type, the main task is to give money to fund a project. Depending on the concrete type of crowdfunding, the amount of money could vary.
5. Crowdopinion. In this case, the task is to gather user opinions about a particular issue or product through different expresions (votes, comments, tags or even sale of shares).

Obviously, each type of initiative implies an interaction with different characteristics between the crowd and the crowdsourcer. These differences make reference, for example, to the required level of interaction, the most suitable reward, the factors that will make a crowdworker get more and more involved, etc.

On the other hand, it is important to point out that the contribution of the people that constitute the crowd is not accepted automatically. On the contrary, there is an assessment process on the part of the crowdsourcer leading to the acceptance or rejection of that contribution. A contribution could be rejected because of low-quality, a requirement missing, late submission, etc.

2.2 *Elements for a Proper Functioning of Crowdsourcing*

Although the eight elements suggested by Estellés-Arolas and González-Ladrón-de-Guevara [11] allow identifying a crowdsourcing initiative, it is true that some of them are more important than others when it comes to achieving success: specifically, the crowd, the technology infrastructure and the rewards. These three elements are also essential in game theory. Other authors enlarge this list by adding, for example, a reputation system o quality assurance [33].

2.2.1 The Medium and the Crowd

The technology infrastructure is so important because crowdsourcing initiatives take place on the Internet, and it is there where interactions happen that lead to an outcome. As Bigham et al. [2] claim, among other things, the crowd will contribute only if the interface guides them in usable and meaningful ways. Each type of platform or environment will facilitate a different interaction, which is why its design is essential in order to obtain the expected final outcome.

On the other hand, as it will serve as a medium of interaction, crowdsourcing initiatives will require a social-technical system beyond the mere infrastructure that can mediate between that infrastructure and the human agent.

Regarding the crowd, it is true that different initiatives will require a different crowd configuration. Certain initiatives will require a more diverse crowd, others a more specialized one, etc. The required crowd size will also vary according to the task: there will be some tasks requiring a crowd of only few people (e.g., the translation of a brief text) and others will require thousands of individuals (e.g., a crowdfunding campaign). One thing is clear – if there is no collectivity, there is no collective intelligence.

On the other hand, the crowd tends to arise around a community of users. On some occasions, those communities gather around a brand or a product (for example Doritos), whereas in other cases they are generated around crowdsourcing platforms suggesting specific types of tasks (e.g. Threadless) [4].

2.2.2 The Rewards

Finally, there is the essential element of the reward, which is the *incentivization* [21]. It refers to those factors motivating the members of a crowd not only to contribute and participate in a initiative [2], but also to do it with an adequate quality. As Ghosh [14] claims, when it comes to the issue of the incentives, there are two key elements to be taken into consideration: firstly, identifying the cost-benefit ratio for each individual of the crowd that is expected to take part in the crowdsourcing initiative; secondly, deciding how and to what extent the contributions of the crowd will be rewarded.

The first element can be directly identified with one of the genes of the collective intelligence suggested by Malone et al. [23] - *why* one should participate in a collective intelligence initiative, crowdsourcing in this case. In other words, what does a reward represent for the individuals of the crowd?

The answer to this question is fundamental, to the point that Malone et al. [23] claim that only adequate incentives can make a collective intelligence initiative arise. These authors also identify three fundamental elements (money, love and glory) that answer that question. But others carry out a more detailed analysis. When it comes to rewards that can motivate the crowd, two types can be distinguished [14, 21]:

- Intrinsic motivators, that arise from internal factors as fun, interest, satisfaction of benefiting a cause, need for social contact, enjoyment or identification with a community, for example.
- Extrinsic motivators, that arise from external factors. In this case, there can be a distinction between purely extrinsic motivators as awards, money or external obligation, and other extrinsic social motivators such as attention, reputation or status.

With regard to which could be the appropriate incentive, it will depend on a series of factors such as the main goal of the task that is being performed, who carries out that task, or even other socio-cultural factors that must be taken into consideration. Jiang et al. [20], for example, carried out a research about which were the best valued incentives for a group of people that had been carrying out micro-tasks in the platform Amazon Mechanical Turk. The result was that Northamerican participants valued the economic aspect the most, whereas the Indians valued to a greater extent the knowledge and skill-related benefits.

The second element refers to how to reward the performance of the tasks proposed in the crowdsourcing initiative. Obviously, within the process of taking an individual decision about whether to take part or not in a crowdsourcing initiative, there is an analysis between the benefit (the reward) that will be obtained and the cost that needs to be incurred in order to participate. This cost will surely imply an appreciation of the time that should be spent in order to carry out that task and the associated effort, although depending on the intitiative there could be other factors at stake. In the case of crowdfunding there is an economic cost, in the case of the crowdcontest there is a cost related to the risk of not obtaining the reward, etc.

3 Crowdsourcing and Game Theory

As it has already been pointed out at the beginning of this chapter, game theory is an appropriate vehicle to express collective intelligence initiatives. After all, regardless the way it is managed, a collective intelligence initiative is based on the interaction of two or more agents.

In fact, different authors have already used game theory in different collective intelligence applications. Ho et al. [16], for example, study a game for semantic annotation of images called PhotoSlap, and provide a game-theoretic analysis of it. Jain and Parkes [19], on their part, analyse from the perspective of game theory different GWAPs (Game With A Purpose) like ESP or FoldIt. These are games in which players generate useful data as a by-product of play [32]. Ho and Chen [15], in this same field, found out that, for example, a simple equilibrium analysis can explain the differences regarding the content generated by different users.

Also crowdsourcing, as a way to manage collective intelligence, has attracted attention from the game theory perspective. In this case, the sudy has been focused mainly on crowdcontest-like initiatives. One of the first papers was that of [7], who focused on the relationship between the incentives and the participation in these types of initiatives modelling these contests as all-pay auctions with incomplete information. Narodirskiy et al. [26] were interested in the relationship between the productivity increasement and the effects of malicious behaviour on the part of the crowd. Other authors have studied other particularities of this type of initiatives, like [5, 33, 34].

Crowdfunding, one of the most booming crowdsourcing types, has attracted the attention of Turi et al. [30, 31] by its relationship with self-enforcing protocols like co-utility. These authors have also made a first approach to the game-theoretic model of other kind of crowdsourcing initiatives.

In spite of these papers, it would be extremely useful to elaborate a general model that could enable the design and analysis of the interactions that occur in collective intelligence initiatives (regardless of their type) as already suggested by Jain and Parkes [19].

From the game-theoretic point of view, any interaction or game can be identified on the basis of a series of elements [22]. A finite game in normal form is a tuple *(N, A, u)*, where:

- N is a finite set of n agents, indexed by i;
- $A = A_1 \times A_2 \times \ldots \times A_n$, where A_i is a finite set of actions available to agent i;
- $u = (u_1, \ldots, u_n)$ where $u_i : A \to \mathbb{R}$ is a real-valued utility (or payoff) function for agent i.

The agents want to obtain – through a strategy- a benefit, and, since they are rational agents, they look for an optimal strategy: a strategy that maximizes the agent's expected payoff.

Nevertheless, this payoff does not only depend on an agent's choices, but on the choices of others. When every agent plays his best strategy (the series of action choices maximizing his payoff given the choices of the others), an equilibrium is reached. When this situation occurs, nobody has an incentive to deviate from his strategy. If, in addition, the result obtained is the best one among the possible results, it is said that the outcome is Pareto-optimal.

By applying the same pattern to the crowdsourcing, and by taking into account the eight elements identified by Estellés-Arolas and González-Ladrón-de-Guevara [11], the following correspondences can be identified:

- A finite set of agents N. This set of agents includes all the persons involved in the crowdsourcing initiative: the crowdsourcer and each person in the crowd (the crowdworkers).
- The actions for persons in the crowd are to perform or not to perform the task or tasks proposed by the crowdsourcer; the actions for the crowdsourcer are either to accept or reject the performed tasks.
- The utility or payoff for persons in the crowd is the reward being offered; the utility for the crowdsourcer is the benefit that the crowdsourcer gets from the realization of the task.

There are other elements that should be considered for a complete game-theoretic modelling of crowdsourcing: the cost of performing the task, the risk the task implies, the extent to which the individuals of the crowd find the task appealing, the quality of the contribution, the feedback, etc. For the purpose of this chapter, the above simple schema will be enough.

4 Co-utility in crowdsourcing

The importance of co-utile protocols is that: (i) they are voluntarily adhered to by rational agents to maximize their own benefit; (ii) by selfishly following the protocol, an agent also helps other agents. That is to say, the selfish behaviour of an agent is the most beneficial for the rest.

The use of this type of protocols in crowdsourcing initiatives occurs in a natural way, at least initially. Except in those cases in which, for whatever reasons, the objective of the participants is to sabotage the initiative (because of professional or personal purposes), the matching between a crowdsourcer and a crowdworker always entails mutual benefis and implies a natural synergy between both agents.

As it is claimed by Domingo-Ferrer et al. [8, 9], for a protocol to be considered co-utile, it must satisfy the following conditions:

1. It is self-enforcing (that is, following it is an equilibrium for the agents);
2. It is Pareto-optimal;

The above means that, in a crowdsourcing initiative, the protocol the enables a successful achievement must be the protocol that people want to follow voluntarily because there is no other protocol, including the no-participation, which may offer a better outcome.

The graphical representation of Fig. 1 depicts crowdsourcing from the crowdworker's perspective. In this scenario there is only one interaction (1 crowdworker vs the crowdsourcer), unlike in the scenario from the crowdsourcer's viewpoint, which would confront the crowdsourcer to many crowdworkers. In this sense, it is worth noting that, although collective intelligence, and in particular crowdsourcing, are based on the collectivity, the final outcome is still the sum or integration of the individual results based on individual interactions and decisions. The person, after all, is the fundamental unit.

In the interaction of Fig. 1 different elements can be identified:

- Two agents, being agent A the crowdworker and agent B the crowdsourcer;
- A utility identified by u_i;

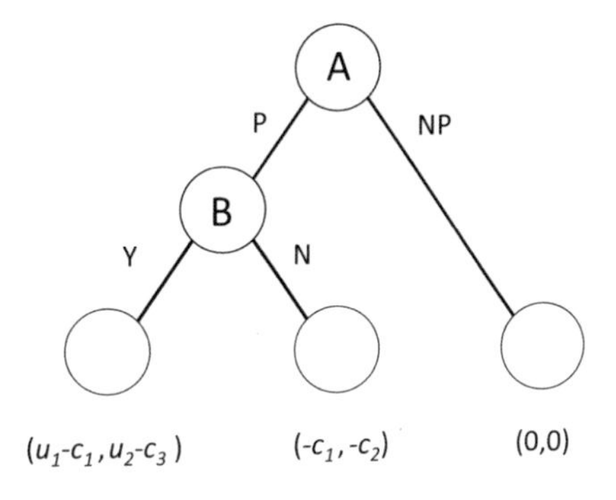

Fig. 1 Crowdsourcing interaction between the crowdworker (agent A) and the crowdsourcer (agent B), expressed in extensive form

- A cost of each action c_i;
- Three different possible protocols: *(P,Y)*, *(P,N)* and $(NP, \emptyset)$;
- The payoffs associated with each protocol.

In this scenario, upon receiving the open call agent A has the options of *Participating (P)* or *Not Participating (NP)*. If he decides to reject the proposed task and not participate, the chosen protocol is $(NP, \emptyset)$ and the outcome is 0 for both players.

If agent A decides to get involved in the initiative and carry out the task, it is up to agent B to make the following decision regarding the result of the task: *accept* (Y) or *reject* (N) the delivery.

If agent B rejects the delivered task, the chosen protocol is (P, N) and agent$A$ incurs a negative payoff because of the cost of time and resources used in performing the task. This cost is represented by c_1. Agent B will also get a negative payoff, of less quantity though. This is because checking the delivered contributions represents a small cost for the crowdsourcer. But this small cost multiplied by the size of the crowd can make an initiative non-profitable; hence, the cost of checking must be taken into account.

In case agent B decides to accept the delivered task, the chosen protocol is (P, Y). In this protocol, A obtains the promised reward (u_1) minus the cost of performing the task, previously identified as c_1; agent B obtains the utility of the correctly completed task (u_2) minus a cost c_3. This cost consists in the cost c_2–to check the delivered contribution–, plus the cost of incorporating and integrating that contribution to the final solution.

For crowdsourcing to make sense, it is necessary that $u_1 > c_1$ and $u_2 > c_3$, which means that the payoffs obtained must be positive. A negative payoff would be especially disastrous for the crowdsourcer, because it would be multiplied by the crowd size.

If the requirements in the previous paragraph are met, protocol (P, Y) is the only one that provides positive payoffs to both players. Hence, it is a co-utile protocol.

5 Conclusions

The widespread use of the Internet and its integration into people's lives are yielding a greater number of interactions that should be designed and modeled to be as efficient as possible, to give the best result for all the agents involved.

In the case of crowdsourcing, and collective intelligence in general, this modelling is even more important but also more complex. That is the reason why self-enforcing protocols, and in particular co-utile protocols, can help simplify the interactions and ensure a smooth functioning.

It is true that the best situation is the one in which a protocol is naturally co-utile (without added incentives), but the reality is that this does not usually happen. In those situations, many of which appear in crowdsourcing initiatives, the protocols can be made co-utile by adding the right incentives [31].

Although different authors have made various proposals about the game-theoretical modelling of crowdsourcing, including the recent co-utility concept, those are limited to specific cases of crowdsourcing. It would be useful to develop a complete model of crowdsourcing encompassing all its manifestations (crowdcontest, crowdvoting, crowdcreation, etc.) and all the variables involved.

References

1. Banks, J., Potts, J.: Co-creating games: a co-evolutionary analysis. New Media Soc. **12**(2):253–270.(2010)
2. Bigham, J.P., Bernstein, M.S., Adar, E.: Human-computer interaction and collective intelligence. In: Malone, W., Bernstein, M.S. (eds.) Handbook of Collective Intelligence. MIT Press, Cambridge (2015)
3. Brabham, D.C.: Crowdsourcing as a model for problem solving an introduction and cases. Converg. Int. J. Res. New Media Technol. **14**(1):75–90 (2008)
4. Brabham, D.C.: Crowdsourcing. MIT Press, Cambridge (2013)
5. Chawla, S., Hartline, J.D., Sivan, B.: Optimal crowdsourcing contests. Games Econom. Behav. (2015)
6. Chesbrough, H.W.: Open Innovation: The New Imperative for Creating and Profiting from Technology. Harvard Business Press, Verlag (2006). (Colman, A.M.: Game Theory and Its Applications: In the Social and Biological Sciences. Psychology Press, Boston (2013))
7. DiPalantino, D., Vojnovic, M.: Crowdsourcing and all-pay auctions. In: Proceedings of the 10th ACM Conference on Electronic Commerce, pp. 119–128. ACM, New York (2009)
8. Domingo-Ferrer, J., Martínez, S., Sánchez, D., Soria-Comas, J.: Co-utility: self-enforcing protocols for the mutual benefit of participant. Eng. Appl. Artif. Intell. **59**, 148–158 (2017)
9. Domingo-Ferrer, J., Sánchez, D., Soria-Comas, J.: Co-utility: self-enforcing collaborative protocols with mutual help. Prog. Artif. Intell. **5**(2):105–110 (2016)
10. Dugatkin, L.A., Reeve, H.K. (eds.): Game Theory and Animal Behavior. Oxford University Press, Oxford (2000)
11. Estellés-Arolas, E., González-Ladrón-de-Guevara, F.: Towards an integrated crowdsourcing definition. J. Inf. Sci. **38**(2), 189–200 (2012)
12. Estellés-Arolas, E., Navarro-Giner, R., González-Ladrón-de-Guevara, F.: Crowdsourcing fundamentals: definition and typology. In: Advances in Crowdsourcing, pp. 33–48. Springer International Publishing, Berlin (2015)
13. Geiger, D., Seedorf, S., Schulze, T., Nickerson, R.C., Schader, M.: Managing the crowd: towards a taxonomy of crowdsourcing processes. In: Proceedings of the Seventeenth Americas Conference on Information Systems, Detroit, Michigan, 4–7 Aug 2011
14. Ghosh, A.: Game theory and incentives in human computation systems. In: Handbook of Human Computation, pp. 725–742. Springer, New York (2013)
15. Ho, C.J., Chen, K.T.: On formal models for social verification. In: Proceedings of the ACM SIGKDD Workshop on Human Computation, pp. 62–69. ACM, Boston (2009)
16. Ho, C.J., Chang, T.H., Hsu, J.Y.J.: Photoslap: a multi-player online game for semantic annotation. In: Proceedings of the National Conference on Artificial Intelligence, vol. 22, No. 2, p. 1359 (2007)
17. Howe, J.: The rise of crowdsourcing. Wired **14**(6), 1–4 (2006)
18. Howe, J.: Crowdsourcing: How the Power of the Crowd is Driving the Future of Business. Random House, New York (2008)
19. Jain, S., Parkes, D.C.: The role of game theory in human computation systems. In: Proceedings of the ACM SIGKDD Workshop on Human Computation, pp. 58–61. ACM, Boston (2009)

20. Jiang, L., Wagner, C., Nardi, B.: Not just in it for the money: a qualitative investigation of workers' perceived benefits of micro-task crowdsourcing. In: 2015 48th Hawaii International Conference on System Sciences (HICSS), pp. 773–782. IEEE, New York (2015)
21. Larson, M., Cremonesi, P., Said, A., Tikk, D., Shi, Y., Karatzoglou, A.: Activating the crowd: exploiting user-item reciprocity for recommendation. In: The First Workshop on Crowdsourcing and Human Computation for Recommender Systems, ACM Conference Series on Recommender Systems, ACM RECSYS (2013)
22. Leyton-Brown, K., Shoham, Y.: Essentials of game theory: a concise multidisciplinary introduction. Synth. Lect. Artif. Intell. Mach. Learn. **2**(1),1–88 (2008)
23. Malone, T.W., Laubacher, R., Dellarocas, C.: The collective intelligence genome. MIT Sloan Manag. Rev. **51**(3):21 (2010)
24. Manshaei, M.H., Zhu, Q., Alpcan, T., Bacşar, T., Hubaux, J.P.: Game theory meets network security and privacy. ACM Comput. Surv. **45**(3),25 (2013)
25. Muggy, L., Heier Stamm, J.L.: Game theory applications in humanitarian operations: a review. J. Humanit. Logist. Supply Chain Manag. **4**(1),4–23 (2014)
26. Naroditskiy, V., Jennings, N.R., Van Hentenryck, P., Cebrian, M.: Crowdsourcing contest dilemma. J. R. Soc. Interface **11**(99) (2014)
27. Oomen, J. Aroyo, L.: Crowdsourcing in the cultural heritage domain: opportunities and challenges. In: Proceedings of 5th International Conference on Communities Technologies – C&T. Queensland University of Technology, Brisbane, Australia, 29 June–2 July 2011
28. O'Reilly, T.: What is Web 2.0: Design patterns and business models for the next generation of software. Commun. Strat. **1**:17 (2007)
29. Sambuli, N., Crandall, A., Costello, P., Orwa, C.: Viability, verification, validity: 3Vs of crowdsourcing. iHub Research (2013)
30. Turi, A.N., Domingo-Ferrer, J., Sánchez, D., Osmani, D.: Co-utility: conciliating individual freedom and common good in the crowd based business model. In: 2015 IEEE 12th International Conference on e-Business Engineering (ICEBE), (pp. 62–67) (2015)
31. Turi, A.N., Domingo-Ferrer, J., Sánchez, D., Osmani, D.: A co-utility approach to the mesh economy: the crowd-based business model. Rev. Manag. Sci. **1–32** (2016)
32. von Ahn, L.: Games with a purpose. Computer **39**(92–94), 2006 (2006)
33. Wu, W., Tsai, W.T., Li, W.: An evaluation framework for software crowdsourcing. Front. Comput. Sci. **7**(5),694–709 (2013)
34. Wu, W., Tsai, W.T., Hu, Z., Wu, Y.: Towards a game theoretical model for software crowdsourcing processes. In: Crowdsourcing, pp. 143–161. Springer, Berlin, Heidelberg (2015)

Problems in the Undertakings of the Collaborative Economy: Co-utile Solutions

Abeba Nigussie Turi, Josep Domingo-Ferrer and David Sánchez

Abstract In the current era of the digital economy, the collaborative economic system, being a successor of the capitalist system, has led to a greater wave of sharing of scarce resources and to the creation of interdependent networks of rational beings. In spite of its promising features of abundant liquidity and efficient utilization of the underutilized scarce economic resources, the system is still in its early stages of development and suffers from some serious problems that hamper its deployment. In order to unlock the potential of this economic system, there is a need for conciliating the individual freedom of economic agents and the common good of the information society encompassing it. In this chapter, we tackle this issue by relying on the recently minted concept of co-utility, which refers to a self-enforcing and mutually beneficial interaction among self-interested agents. By means of co-utility, we aim at mending some of the fractures underlying the collaborative economy due to key hindering factors for potential collaboration. Further, the potential collaborations within each of the use cases presented in this text (i.e., the crowd-based business models and the P2P online lending market) are fostered by co-utile solutions, arising either naturally or artificially through incentive schemes.

1 Introduction

The collaborative economy (also known as mesh or sharing economy) refers to technology-oriented supply-demand facilitation, with an on-demand extension of the concept to the previously underutilized resources [12]. This has resulted in an efficient utilization of scarce resources, which is the main motivation underlying the

A.N. Turi (✉) · J. Domingo-Ferrer · D. Sánchez
UNESCO Chair in Data Privacy, Department of Computer Science and Mathematics, Universitat Rovira i Virgili, Av. Països Catalans 26, E-43007 Tarragona, Catalonia, Spain
e-mail: abebanigussie.turi@fundacio.urv.cat

J. Domingo-Ferrer
e-mail: josep.domingo@urv.cat

D. Sánchez
e-mail: david.sanchez@urv.cat

J. Domingo-Ferrer and D. Sánchez (eds.), *Co-utility*, Studies in Systems, Decision and Control 110, DOI 10.1007/978-3-319-60234-9_12

economic theory. In its raw sense, the core principles assumed to govern the system are: collaboration, empowerment, transparency, humanity and altruistic sharing for the common well-being.

Rifkin [26] argues that there is a paradox in the capitalist system and that the invisible hand that has been responsible for the capitalism's success has led to a new successor paradigm called the collaborative economy. Yet, some of the new business models we came across with this trend have elements of altruism and rent combined. For the collaborative system not to collapse and destroy itself, the values derived should rest on the principles of mutual help and ethics. These are precisely the values underlying to co-utility, a novel concept for designing interaction protocols among agents that conciliate their selfish and rational choices with societal welfare (see [7, 8], and the first chapter of this book).

By proposing co-utile solutions for the collaborative economy, our aim here is to foster its applications and its core principles in all potential areas, with a demand-led lens like all novel innovations. Specifically, co-utile protocols, being self-enforcing and mutually beneficial, contribute to making transactions efficient by suppressing sub-optimal interactions.

In this chapter, we will be focusing mainly on the crowdsourcing and FinTech markets in general, including crowdfunding, P2P lending and cryptocurrencies.

2 The Collaborative Economy: An Overview

The collaborative economy is an economic system characterized by the underlying key economic features of sharing, leasing, swapping, selling, buying, lending, giving and bartering. With these key transactional features, it has the potential to unlock the idle capacity in the utilization of the scarce economic resources mainly using, but not limited to, the Internet [12]. Financial technologies including the business lines of peer-to-peer (P2P) transactions, crowdfunding and crowdsourcing, innovation and educational marketplaces, are some of the common structures of this economy. The collaborative economy paradigm revolves around the core principles of collaboration, empowerment, transparency, humanity and altruistic sharing for the common well-being, which results in efficiency with no hyper-consumption. The emergence of irreversible consumer behavior is one of the catalysts for a widespread replication of this system all across the globe. By irreversible consumer behavior we mean a preference shift of the digital society to a new form of utilization of goods and services instead of the traditional form of consumption.

According to [26], capitalism has led to its own mutation into a new collaborative economic system that he calls the 'zero marginal cost society'. However, some scholars argue that the collaborative economy, which has initially manifested itself in the form of sharing economy, has more tendency to be pure capitalism. For example, [24] proposes networks with members' co-ownership rather than operating in the form of monopolistic platforms like Uber. His argument sounds like a call for a co-utile interaction, which we will introduce in the following section. A prominent example of a successful co-ownership model in the real world is the John Lewis Partnership, a UK-based retail company co-owned by the employees [28].

The collaborative economy is characterized by abundant liquidity. PricewaterhouseCoopers in 2015 estimated that, by the year 2025, transactions under this economy are expected to generate about $335 billion at a global level. Despite its rich liquidity for the players, the collaborative economy has posed many challenges that prevent traditional players from embracing it. The new business models of this economic system are disrupting the incumbent customer base of traditional sectors and also making their business models and services obsolete. On the other hand, with the new business models of collaborative economy, new risks arise. These new models are exposed to new challenges and uncertainties unique to their individual set-up. This is mainly due to the dynamics and hasty evolution of the business models to cope with the new regulatory and legal codes and latest technology-based competition from their players. Furthermore, this economy poses a challenge to the governments, for which enacting new rules and regulations that cope with the dynamics of the collaborative business models is hard. This will also disrupt the government's revenue with a fundamental transformation of the market players joining the new business models before new rules are enacted. Another important issue is the mistrust between the players within the system itself. This results from the uncertainties and asymmetric information that exist in most of the transactions of the collaborative economy models. Hence, in this system, reputation is a highly valued asset, and it is regarded as the currency and capital of the collaborative economy (see [2, 11]).

3 Co-utility

Co-utility refers to a philosophy of economic and social norms based on mutually beneficial interactions. It explains mutuality in the framework of the social rational choice theory, in which an agent taking part in any form of interaction is bound to a self-rational choice which is in line with the rational choice of another agent it interacts with. Formally defined, an outcome for any game is co-utile, if and only if (1) it is equilibrium of the game, (2) it is a Pareto-optimal outcome and (3) playing the game gives a strictly higher payoff to all the players involved than not playing the game [7, 8].

Furthermore, it is worth mentioning the following key points about the co-utility theory.

(i) Players involved in any co-utile interaction can have either symmetric or asymmetric goals. For example, in the case of a P2P file sharing network, the goals of all the agents involved in this network are the same. But if we take P2P online lending, the goal of a borrower is soliciting finance, whereas that of a lender is making profit at a given interest rate.

(ii) A proportional flow of values in a co-utile interaction is a necessary condition in order to keep a positive utility to all the agents involved, yet this flow of values does not need to be equal.

(iii) Co-utility can arise either naturally or artificially through incentive mechanisms. Here, it is important to consider the timing of the payoffs to any co-utile games attained, i.e. immediate or future expected payoffs. This will help capture the dynamics of co-utile interactions between rational agents, especially when designing artificial incentives that will guarantee co-utility in co-utility amenable games. In some cases, an agent may do a favor to another because the other has favored this agent in the past or is expected to do so in the future. Hence, in addition to spontaneous interactions, the co-utility framework can be designed to consider the dynamics with time.

(iv) As stated in the formal definition of co-utility, co-utile interaction results in superior payoff to all agents involved over the next best alternative interaction. Hence, the efficiency of any co-utile game can be measured by quantifying the intrinsic efficiency advantage of co-utile interactions over the other options.

(v) The basic requirements for the fulfillment of an efficient co-utile interaction are collaborative demand (i.e., the ability and willingness of the parties involved in the co-utile game), commitment and trust between involved parties, a common platform for the interaction, and a post-transaction positive utility.

Co-utility is also important in the public relations and interactions of an organization with its customers and stakeholders. For instance, a fair market price service provision by a supplier, which neither undercuts the supplier's profitability nor passes on unnecessary cost increases to the customers, is mutually beneficial both for the sustainability of the business suppliers and for achieving a cost-effective access to the company's services by the customers. Hence, co-utility as a business strategy can deliver sustainable profitability of the business while benefiting the customers. One example of this can be a business strategy followed by Mars Inc., an American global manufacturer of confectionery, pet food, chocolate, candy and other food products. In 1998, when Russia defaulted on its foreign debt, in order to keep their customer base in this country the company designed a lifeline offer of a product without payment until it was resold, which averted bankruptcy and allowed continuing their operations even during this uncertain period.

In a more general and aggregate sense, co-utility may also apply to the two-way democratic communication between the government and the citizens, whose interaction is based on the principles of mutuality, empathy, trust and solidarity. This communication is done through a two-way symmetric communication, characterized by negotiation, conflict resolution and respect. One key type of government interaction with the citizens is carried out through national policy development. For example, in the policy of holding a low inflation, there exists a dynamic inconsistency arising from people's rational expectation on the expansionary/contractionary monetary policy measures taken by the government in response to a low/high expected inflation [27]. Two of the models proposed under this scenario are the reputation and delegation model.

In the reputation model, policy-makers with more than a two-period office stay (i.e. more than two terms) will aim at either maximizing wealth or fighting inflation. A policy-maker can "cheat", because people's expectation (and trust) can be affected by the policymaker's decision, which is the unrevealed type of the policy maker. The reputation model implies that uncertainty about the characteristics of policy-makers reduces inflation. If a policy-maker does not cheat, expectations for inflation will be lower. If it cheats, the reputation is damaged and expectations for inflation will be higher.

On the other hand, a model of delegation delegates the monetary policy to an independent inflation-averse institution. The problem with this model lies in its limitation of the risk aversion principle, for example, during economic shocks. In such case, the implementation of a stabilization policy is limited because the underlying assumption of linear social welfare is too strong while, on the other hand, monetary policy fluctuations also result in fluctuations in consumption and fluctuations in working hours, which result in disutility for the workers. Hence, lags of policy effects, the existing economic state and uncertainty about the future limit the realization of a stabilization policy. Because a stable economy hosts and pulls healthy investments, collaborative and mutually inclusive acts, both by the government and by the citizens, are needed. In this regard, the notion of co-utility can play an important role in the interaction between the government and the citizens.

Our main aim w.r.t. co-utility is to foster the application of the innovative models of collaborative economy and core principles to all potential areas, with a demand-led lens like the replications of all previous novel innovations of the universe. In the next section, we present some of the business models of the collaborative economy that are co-utility-amenable by design and discuss how co-utile interactions may be designed.

4 Co-utility-amenable Models of the Collaborative Economy

This section approaches the collaborative economy by introducing the concept of co-utility into the existing business models, with the aim of enhancing their efficiency and the social welfare in general. These business models, which highly rely on the ICTs, fall under the umbrella of the digital economy (aka Internet or web economy). We start with the crowdsourcing business model, followed by the two main FinTech funding models (crowdfunding and P2P online lending) and last, but not least, cryptocurrencies.

4.1 Crowdsourcing

Crowdsourcing is a method for outsourcing a task to an anonymous crowd. Any individual worker taking part in the crowdsourcing market makes a choice based on a cost-benefit analysis of the participation. There exists a co-utile interaction in this market as long as the goals of requester and worker are complementary and the qualification type of the worker matches the task. Let u_i be the utility function of a worker in the crowdsourcing market. Then,

$$u_i(\mathcal{T}, e) = f_i(\mathcal{T})\left[\alpha_i r_i(e) - c_i(\mathcal{T}, e)\right], \tag{1}$$

where $f_i(\mathcal{T})$ is a binary function specifying whether or not task $\mathcal{T}$ matches worker i's interest and qualifications (ability); $f_i(\mathcal{T}) = 0$ implies worker i is not interested or not qualified to perform the task; $f_i(\mathcal{T}) = 1$ implies worker i is qualified and willing to exert effort towards this specific task; e is the effort level to perform the task; $r_i(e)$ is the expected reward to worker i for effort e; α_i is a weight variable reflecting how much the individual values the reward; and $c_i(\mathcal{T}, e)$ is the task-specific cost of effort to worker i (e.g., it can take the form of time devoted to reading, understanding and performing the task, expenses incurred to perform the task, etc.). If $c_i(\mathcal{T}, 0) = 0$, no effort is then exerted and, for the worker, there is no cost associated to this task. There is also a trade-off between labor and leisure, l, in the participation decision of a worker. This calls for a reasonably satisfactory per-task pay that outweighs the utility of leisure, $u_i(l)$. A rational worker i supplies labor in a crowdsourcing market if $u_i(\mathcal{T}, e) > 0$ and $u_i(\mathcal{T}, e) > u_i(l)$.

Similarly, a rational requester in a crowdsourcing market will choose crowdsourcing over any other way of outsourcing his task based on the expected return. The utility function of requester j, u_j, in a monetary reward-based crowdsourcing depends on the minimum threshold it expects to gain from the task accomplishment ($\mu(\mathcal{T})$), and total output of the assigned task, y. It is defined as:

$$u_j\left(y, w_j, Crd, \mu_j, \mathcal{T}\right) = \alpha_j(y - w_j Crd - \mu_j(\mathcal{T})), \tag{2}$$

where α_j is a weight variable reflecting how much the requester values the overall gain from the output; Crd is the crowd labor supply; and w_j is the per-task pay offered by requester j. A rational requester who wants a task $\mathcal{T}$ to be performed will post the task to the anonymous crowd only if $u_j\left(y, w_j, Crd, \mu_j, \mathcal{T}\right) > 0$.

Both requesters and workers involved in this market maximize their respective utilities and collaborate for their best interest. Therefore, the crowdsourcing market is a potentially co-utile market with the natural incentives underlying the market. Yet, like many other collaborative economy models, this business model suffers from some limitations. The main one is that this model is of limited applicability for any type of task outsourcing, for example, those tasks which require privacy. Another key problem is that monetary rewards to workers are mostly very low, which will have a macroeconomic effect.

4.2 FinTech in a Broader View

Financial technology, abbreviated as FinTech, refers to technology-oriented business lines of the digital economy. FinTech companies range from crowdfunding and peer-to-peer lending to algorithmic asset management and facilitation of investments, financial planning, portfolio management, thematic investing, payments, data collection, credit scoring, education lending, digital currency, exchanges, working capital management, cyber security and quantum computing. Lending Club, Kickstarter, Prosper, Betterment, Xoom, 2iQ Research, ZestFinance, Coinbase, SecondMarket, Tesorio, iDGate, Motif Investing, Stripe, and Square are some examples of marketplaces in this industry. In addition, already-established IT companies like Facebook Apple Pay, Android Pay, and Google Wallet are also actively involved in the money-transfer markets and mobile payments. This wave of technological penetration in the financial industry has created a significant pressure to the traditional incumbent players in the market. Tech startups come up with an improved artificial intelligence, automation, algorithmic decision-making, cost-effective data mining and processing, and a convenient and efficient speed of operation while capturing the low-end customers' preference. According to a report by McKinsey & Co. [19], traditional banks could lose about 60% of retail profit to tech startups in the coming few years.

Unlike the traditional form of business models, each individual user is a co-originator of value in this new system. Hence, we see that there is a space for co-utility underlying these business models. In order to elaborate these in more detail, we present some FinTech sub-industries and business lines which are co-utility-amenable by nature.

4.2.1 Investment Crowdfunding

Crowdfunding is a fund-raising business model that depends on sourcing finances from a given network of crowd, mainly through the Internet. The key players in this market are the project initiator/requestor, the crowd, and an intermediary platform that facilitates a link in the network. It takes the form of either investment crowdfunding including debt-based, equity-based, profit-sharing or hybrids of these, or donation crowdfunding, also called reward-based crowdfunding. Crowdfunding is an example of a co-utility amenable game. The market can be modeled using a financial market imperfection model accounting for the information asymmetry and risk-averse behavior of the investors and project initiators (see [27, 31]). The expected return to a crowdfunding investor i can be defined as:

$$E(R_i) = \pi x_i D + (1-\pi)[-(1+r)C_i - \alpha_i], \tag{3}$$

where $\pi = (\gamma^* - \mathrm{D})/\gamma^*$ is the probability at which the entrepreneur rewards the investor with amount D as per the debt contract and takes the surplus; x_i is the proportion of investor i's investment in the project, C_i is invested amount and r is the random interest rate. $\sum_{i=1}^{n} x_i = 1$ and $D_i = x_i D$. By solving for D, the optimal debt contract at which the return based investment decision takes place is at $D = \frac{\gamma* - (1+r)C_i - \alpha_i}{2x_i}$. Hence, given a negative exponential utility function of investor i: $u(R_i) = -e^{-\beta_i R_i}$, where $\beta_i > 0$ is the risk-aversion factor for investor i, a rational investor i invests if and only if $u_i > \quad 0$ and $u_i(R_i^*) \geq u_i\left((1+r)\,C_i\right)$. A rational entrepreneur j's investment decision also depends on the net expected return from its project given as:

$$E(R_j) = \gamma - [(1+r)C + \alpha], \tag{4}$$

where γ is the expected output of the project. Given a face value of the project v, entrepreneur j broadcasts the project for crowdfunding if $\gamma - [(1+r)C + \alpha] > v$, given that the required rate of return is not more than the optimal rate of return $(1+r)C \leq R^*$. Hence, given the entrepreneur's utility function $u_j(C, r, \gamma, V) = -e^{-\beta(\gamma-(1+r)C-\alpha-V)}$, where $\beta > 0$ is a risk-aversion coefficient for the entrepreneur, it broadcasts its project if $u_j\ (C, r, \gamma, V) > 0$ and $u_j(C, r, \gamma, V) \geq u_j(V)$.

The crowdfunding market is prone to a number of risks due to information asymmetries and uncertainties. Due to the risk underlying this market, it is still in its early stage of development. With the practical problems *underlying the market,* there exists a non-return-based investment decision, which limits the co-utile nature of the market. We can categorize the major problems arising in this market into two main categories: *fear of disclosure effect* (entrepreneurs) and *mistrust effects* (investors). Accounting for the mistrust effect, the investor's utility function can be redefined as

$$u_i\ (R, \Gamma) = \begin{cases} -e^{-\beta_i \Gamma R} & \text{for } \Gamma \neq 0 \\ 0 & \text{for } \Gamma = 0 \end{cases}, \tag{5}$$

where Γ = trust, and $\Gamma \epsilon\ [0, 1]$; $\Gamma = 1$ if investors completely trust the entrepreneur and $\Gamma = 0$ if investors do not trust the entrepreneur. The fear effect on the side of the entrepreneur accounts both for the fear of failure and the fear of disclosure, which implies the risk of copy by competitors, especially for interactive digital media developers and content producers. Accounting for the *fear effect,* the utility function for entrepreneur can be defined as:

$$u_j(C, r, \gamma, V, \mathcal{F}) = \begin{cases} -e^{-\beta F(\gamma-(1+r)C-\alpha-V)} & \text{for } \mathcal{F} \neq 0 \\ 0, & \text{for } \mathcal{F} = 0 \end{cases}, \tag{6}$$

where $\mathcal{F}$implies no-fear and $\mathcal{F} \in [0, 1]$; $\mathcal{F} = 1$ if the entrepreneur can broadcast its project content with no fear and $\mathcal{F} = 0$ if the entrepreneur has maximum fear.

Hence, beyond the return-based investment decision, a co-utile point of collaboration in this market is the Pareto-optimal point at which the optimal level of no-fear and trust is attained. This can be attained through artificial incentive schemes, which can tackle the risks underlying the market. The first mechanism can be enforced through encryption, by securing the private key with a decentralized timestamp for neutralizing the fear effect. The second is to employ an efficient reputation mechanism to resolve the *mistrust effect*. One possible reputation mechanism is the community-based reputation system in which credible projects are signaled by heads of a given group/network, who bear the risk of failure of a project with a sufficient skin in the game [13, 31]. Yet, this mechanism lacks the property of being decentralized and self-enforcing, which prevents it from being co-utile in its operation. Hence, a better-suited option is to use a decentralized co-utile reputation protocol in which reputation scores of all players in the network are computed in a distributed way (see [9] and the discussion on P2P lending below).

4.2.2 Peer-to-peer Lending

Another co-utility amenable FinTech business model is P2P lending. Peer-to-peer lending is a web-based direct borrowing and lending practice between individual agents. This market acts as a substitute for the brokerage in transaction banking functions of the traditional banking. Lending Club and Prosper are some examples of online platforms facilitating the transaction between the P2P lending networks. A relatively higher rate of return in this market attracts potential lenders to invest in P2P loans rather than in any other alternative investments. In the same way, borrowers choose to borrow from the P2P online lending rather than other alternative sources of financing, like banks, mainly due to the relatively lower interest rates, [20]. In addition, borrowers prefer P2P online lending rather than traditional lenders since, in some cases, loans of low credit grade might be approved in P2P lending market, which could be rejected by the traditional banks. Stated formally, the borrower's utility function U_b in this market depends on its residual income after the loan repayment discounted to its present value over the term of the loan, given the urgency on the loan origination at a given time defined as:

$$U_b = U\left(y + \delta\left(t\right)\left(l_0 - P\frac{1-(1+i)^{-n}}{i}\right)\right), \tag{7}$$

where y stands for the borrower's income, the term $\frac{1-(1+i)^{-n}}{i}$ is the present value of a loan amount l_0 with future annuity payments P, n is number of terms of the loan and i is a random interest rate; $\delta(t) \in [0, 1]$ is a weight parameter representing the level of impatience of the borrower to get the loan at a given time t; $\delta(t)$ values close to 1 imply higher patience level of the borrower. A rational borrower makes a loan request that maximizes its utility.

There are two types of lenders in the P2P lending: pure lenders that rely on their own capital for investing in the P2P lending, and lenders with reinvestment, who are those partially or fully relying on a borrowed capital for reinvesting in the same network in order to make profit through arbitrage opportunities in the market, termed as *P2P loan carry trade*. Hence, the utility functions of the lenders in this market depend on their type. The lender's utility depends on the net lending profit. The expected utility function for the pure lender with a loan amount l_0 is given as:

$$u_l = \sum\nolimits_{n=1}^{n} u\left[(i-r)\, l_0 \gamma(t)\, p(t)\right], \tag{8}$$

where i and r are the nominal and real market interest rates, respectively, $\gamma(t)$ is the proportion of the loan paid at time t over the loan period n, and $p(t)$ is the probability of default at time t. The utility function of the borrower with a re-investable borrowing is defined as:

$$u_l' = u(\frac{rL - r'(L-K)}{K} - \frac{\delta^2(\Pi)}{2\beta}\left(\frac{L}{K}\right)^2), \tag{9}$$

where K is the borrower's investable capital, L is the loan amount and r and r' are the random lending and borrowing interest rates, respectively. The lender makes a profit from the spread between r and r' in the loan *P2P loan carry trade.* The utility maximization problem of the lender with a re-investment motive depends on the difference between the total return on the loans, $\frac{rL-r'(L-K)}{K}$, and the total funding cost, $\frac{\delta^2(\Pi)}{2\beta}\left(\frac{L}{K}\right)^2$ which stands for the total risk of this investment as a product of the risk on the loans, $\delta^2(\Pi)$, and lenders leverage, $\left(\frac{L}{K}\right)^2$ (see [4]).

Hence, provided the formulation and discussion above, we argue that the P2P lending market is co-utile as long as it operates well enough and that there is enough finance circulating in the network with a large number of borrowers and lenders. However, the information asymmetry and market uncertainties impose a limit on the efficient and smooth flow of resources within the P2P lending network. In order to avert the problem of mistrust arising in this market, we propose a decentralized co-utile reputation mechanism that allows computing the reputation of each agent in the lending network in a self-enforcing way. In this protocol, local and global reputations of borrowers in the network are computed based on the outcome of direct transactions between borrowers and lenders.

The local reputation of borrower j, with respect to lender i, is the aggregate payoffs that i has gained from the set of direct transactions, Y_{ij} (capturing i's total satisfaction/dissatisfaction over the set Y_{ij}). It is given by $S_{ij} = \sum_{y \in Yij} u(x)_{iy}$, where $u(x)_{iy}$ is the utility of lender i with money return x, in the transaction $y \in Y_{ij}$, performed with a borrower j. The utility from a lending transaction can be redefined as $u(x)_{iy} = 1 - 2/(1 + \exp(x))$, which gives values between $[-1, +1]$ so that satisfaction scores are computed in a uniform fashion; $u(x)_{iy}$ values close to -1 represent dissatisfaction due to default and $u(x)_{iy}$. values close to +1 reflect satisfaction. Following [15]), the local reputation values are normalized to be positive for a potentially non-defaulting borrower and zero otherwise (i.e., for defaulting borrower or a newcomer to the market). A matrix of normalized local reputation values C for the whole lending network is constructed with zero entries for those borrowers and lenders with no interaction. Furthermore, this protocol works under the assumption that there are some pre-trusted agents who work for the well-being of the system, lenders in our case. Hence, lenders are assigned a small positive reputation score of $1/n$, where n is the number of players in the network. Another important assumption under this protocol is that trust is transitive, that is, if lender i trusts borrower j, then she also trusts another borrower k, who is trusted borrower j. This assumption is important for the computation of the global reputation, which is the aggregated normalized local reputation over all the set of direct transactions of the borrower.

The global reputation score g_{ik} of k is estimated as $\hat{g}_{ik} = \sum_j c_{ij} c_{jk}$, where c_{ij} is the normalized local reputation values of k from other agents j who have a direct lending transaction with k, and c_{jk} is the local reputation value of j from the past record of transactions with lender i. The co-utile nature of the protocol lies in the fact that the global reputation of each agent k is computed by one or several other agents (called score managers of k) on behalf of k, which themselves are lending and borrowing peers participating in the same network. Through successive computations, the global reputation scores converge and sum up to one. Thus, the global reputation score for the whole network is the left eigenvector of the normalized local reputation and it is initialized with a pre-defined small positive global reputation score. The protocol is self-enforcing and hence co-utile with interesting features, such as being anonymous, cost-effective and attack-tolerant (see [9, 30] for the details on the construction and computation).

4.2.3 Cryptocurrency and the Blockchain Technology

i. Cryptocurrency

Cryptocurrencies are another example of digital finance models that are co-utility-amenable. They are considered as the virtual game-changers that meet the FinTech business model needs for digital value transfer and exchange. This technology has made possible geographically unbounded, frictionless, secure and cost-effective flows of values across the globe. Cryptocurrencies like Bitcoin (BTC) are peer-to-peer payment systems in which direct transactions between peers are conducted without a central party. Transactions between peers are confirmed by network nodes and recorded in the blockchain (i.e., a public mutually distributed ledger). The blockchain uses cryptology to guarantee trust between transacting peers in a decentralized way. Some traditional banks and financial institutions facing competition from the new FinTech wave are working to use this technology for a cross-border money transfer and trading (e.g. CBA, Australia, which uses the Ripple payments).

Despite their promising features, these decentralized virtual currency systems also have some limitations. Some of these include (1) difficulty in the exchange with other currencies, (2) being a tool for black market operation and tax evasion purposes in some cases, (3) being prone to potential theft attack of private keys, (4) expensive electricity cost for computations and (5) a possible attack by selfish miners/mining pools who could subvert the system for their own benefit [10, 14, 16].

In order to characterize the behaviors of miners in the Bitcoin network, we present the model in the following simplistic way. In its current setting, miners are rewarded based on successfully mining a new block (fixed amount of 25BTC $\approx$ 12500€ at the current rate of exchange, while the correspondence between the BTC and euros is not fixed) and additional 1.33% of the transactions in the added block [16]. Every miner in Bitcoin aims at maximizing the revenue from mining. Hence, the profit function of any miner, m is defined as follows:

$$\pi_m(r, f(\lambda_m), c_m(\lambda_m, t)) = f(\lambda_m) \cdot r - c_m(\lambda_m, t), \tag{10}$$

where r is the reward, λ_m is the miner's relative computational power as compared to other miners in the network, $f(\lambda_m)$ is a binary function of successfully solving the puzzle on time ($f(\lambda_m) = 1$ if the miner successfully mined a new block in time before other miners given its computational power λ_m, and $f(\lambda_m) = 0$ if the miner fails to successfully mine the new block on time), $c_m(\lambda_m, t)$ is the unit cost of mining a new block, which depends on the investment in the computing resources and also the time devoted to solving the puzzle, t is the opportunity cost of mining any other new block. The miner's total profit that he wants to maximize is the sum of profits from all the new blocks he mined. Thus, the miner's profit maximization problem can be stated as:

$$Max \sum_{i=1}^{n} [f(\lambda_m).r - c_m(\lambda_m, t)], \tag{11}$$

where n is the total number of new blocks mined by miner m.

Given the profit maximization problem stated above, miners in the Bitcoin network use various strategies by investing in expensive computational resources or employing a number of strategic attacks against other competent players in the network. In practice, the bitcoin mining game is an asymmetric information game. Rational miners in the Bitcoin network decide on the identification of the block to be mined and when to release or withhold the block [16, 22]. Due to the information asymmetry underlying the game, there can be a number of selfish mining attacks. A game-theoretic analysis by [16] shows that, under both asymmetric and complete information of a mining game, miners with relatively smaller computational power will not deviate from the bitcoin system design, while those with larger computational power will have an incentive to deviate. [10] also assert this magnitude effect by the colluding selfish miners who can generate unfair revenues. Likewise, [6] and [22] showed that a selfish mining pool can make a profit by withholding the block to infiltrate other mining pools in the mining network and consequently make unfair returns. On the other hand, [14] showed that mining pools with a relatively larger computational power are also exposed to a DDoS attacks by the smaller ones aimed at slowing down and weakening the target pools.

Regardless of all these drawbacks, cryptocurrency as a medium of exchange is one example of co-utility amenable game in which encrypted currencies are minted by self-interested miners through a mathematical race that protects the blockchain within the network. With a decentralized setting of the network of miners who keep public records, potentially selfish miners can make the system inefficient. Hence, to make the ecosystem more robust and co-utile, further research work is needed to develop incentive schemes that can hinder malicious miners from subverting the system.

ii. Application of blockchain technology in organizational business processes

Beyond the cryptocurrencies, key features of the blockchain technology that are attractive to the financial institutions and to other transactional networks are: (i) it is decentralized; (ii) it is secured with cryptographic validation of transactions; (iii) it is reasonably efficient; and (iv) transaction records are transparent; see for e.g., [18, 23] on the applications of the blockchain technology to transactional record systems and financial services other than the cryptocurrencies, some of which include securities settlement, currency exchange, supply chain management, trade, P2P transfers, asset registration, correspondent banking, regulatory reporting and AML rules. The technology guarantees efficiency in time because it lets all the stakeholders work in a common dataset and the transaction data are efficiently organized. It also allows easy information exchange with a one-time data entry in a distributed way, without a double and separate record of events, which results in a significant reduction in costs of data reconciliation, checks and transfers. Furthermore, the potential of the blockchain technology to support smart contracts makes it attractive to the financial service sector, because it can be extended with additional features that can fit the purpose of the financial service industry; for example, a permissioned blockchain network with centrally assigned transaction validators (see [29]). The technology facilitates a co-utile business process and organization by allowing rational (selfish) players to interact in a self-enforcing and distributed way.

A specific suggestion in this regard can be the potential of the blockchain technology's application to the financial management and administration industry. A trust company acts as a trustee (someone who acts as a custodian for trusts, estate-oriented services such as guardianship, estate settlement, custodial arrangements, asset management, stock transfer, beneficial ownership registration, etc. and provides audit, tax, consulting, bill pay, check writing, enterprise risk and financial advisory services and other related arrangements) to the other customer enterprises in a business-with-business (BwB) like deals (e.g. Deloitte, Northern Trust, Bessemer Trust etc.). One of the key problems underlying this business sector is the lack of trust on the trust companies, which poses a limit on the efficiency of the service delivered by the financial administration and management companies. As a result, there is a need for development of a collaborative system in which all the stakeholders in this market can act towards the value-creation process and hence minimize the mistrust effect underlying the industry.

Therefore, the blockchain technology with its aforementioned features can help create an efficient financial service industry integrating the BwB interactions of this type in a transparent and well-organized way. In line with this, there is a need to transform the rigid accounting business models to a further digitalization and automation, and to define new potential market niches. This is because, with the newly emerging business models, the trust companies should cope with the existing technological and socio-economic trends. In this regard, we recommend the use of artificial intelligence in the sector [1, 17] and the employment of co-creative and co-utile value creation mechanisms building a secure network of transactions. The trust in the network of financial services can be built by organizing a common platform that provides security and privacy by default. The business activity logs and financial data of any registered client company in the platform are saved in a common database and a potential registered and authorized trust company can access the client's data, which are secured through data protection techniques. Possible data protection methods are data anonymization/splitting (see [3, 5, 25] on the outsourcing of computation without outsourcing control for data in the cloud), homomorphic encryption [21] or privacy preserving data mining. Here, we point out the direction of work in this regard, while specific technical design and implementations are beyond the coverage of this chapter.

5 Concluding Remarks

This chapter introduced the co-utility notion as a mean to mend some of the fractures in the collaborative economy. The main purpose of co-utility is to make the collaborative and sharing principles underlying the collaborative economy self-enforcing. Co-utility in this setting can be enforced naturally or artificially through incentives. Examples of the potential co-utility-amenable games of the collaborative economy include crowdsourcing, crowdfunding, cryptocurrencies and P2P online lending.

In order to solve the key problems of mistrust and fear that appear in the crowdfunding business model, incentive schemes are proposed. These schemes are a community-based reputation scheme and data encryption. Specifically, to tackle the mistrust effect in the peer-to-peer online lending market, a distributed and co-utile reputation protocol is leveraged, so that rational peers rationally cooperate to compute each other's reputation scores. Reputation scores of borrowers are computed based on the outcome of direct transactions. Then the reputation mechanism helps filter credible borrowers based on their respective reputation scores. This reputation mechanism can be used to rate buyers and sellers in the electronic commerce market as well.

Last but not least, we have discussed cryptocurrencies and the blockchain technology underpinning the new trends of the digital economy. The discussion highlights the potential applications of blockchains and distributed ledger technology in the business-to-business transactional networks and proposes a business process design method based on the co-utility notion accounting for the collaborative nature of the system. We further discussed the behaviors of miners in the Bitcoin mining network and the negative utilities arising from the potential attacks. Based on this, we draw a general roadmap in order to achieve co-utility in the system.

Acknowledgments and disclaimer Funding by the Templeton World Charity Foundation (grant TWCF0095/AB60 "CO-UTILITY") is gratefully acknowledged. Also, partial support to this work has been received from the Government of Catalonia (ICREA Acadèmia Prize to J. Domingo-Ferrer and grant 2014 SGR 537), the Spanish Government (projects TIN2014-57364-C2-1-R "SmartGlacis", TIN2015-70054-REDC and TIN2016-80250-R "Sec-MCloud") and the European Commission (projects H2020-644024 "CLARUS" and H2020-700540 "CANVAS"). The authors are with the UNESCO Chair in Data Privacy, but the views in this paper are the authors' own and are not necessarily shared by UNESCO or any of the funding bodies.

References

1. Baldwin, A.A., Brown, C.E., Trinkle, B.S.: Opportunities for artificial intelligence development in the accounting domain: the case for auditing. Intell. Syst. Account. Financ. Manag. **14**(3), 77–86 (2006)
2. Botsman, R., Rogers, R.: What's Mine is Yours: the Rise of Collaborative Consumption. Harper Collins, New York (2010)
3. Calviño, A., Ricci, S., Domingo-Ferrer, J.: Privacy-preserving distribution of statistical computation to a semi-honest multi-cloud. In: 2015 IEEE Conference on Communications and Network Security-CNS 2015, pp. 506–514. IEEE, New York (2015)
4. Carlin, W., Soskice, D.: Macroeconomics: Institutions, Instability, and the Financial System. Oxford University Press, Oxford (2014)
5. Chow, R., Golle, P., Jakobsson, M., Shi, E., Staddon, J., Masuoka, R., Molina, J.: Ontrolling data in the cloud: outsourcing computation without outsourcing control. In: Proceedings of the 2009 ACM Workshop on Cloud Computing Security, pp. 85–90. ACM, Boston (2009)
6. Courtois, N.T., Bahack, L.: On subversive miner strategies and block withholding attack in Bitcoin digital currency. arXiv:1402.1718 (2014)
7. Domingo-Ferrer, J., Martínez, S., Sánchez, D., Soria-Comas, J.: Co-utility: self-enforcing protocols for the mutual benefit of participants. Eng. Appl. Artif. Intell. **59**, 148–158 (2017)
8. Domingo-Ferrer, J., Sánchez, D., Soria-Comas, J.: Co-utility: self-enforcing collaborative protocols with mutual help. Prog. Artif. Intell. **5**(2), 105–110 (2016)
9. Domingo-Ferrer, J., Farràs, O., Martínez, S., Sánchez, D., Soria-Comas, J.: Self-enforcing protocols via co-utile reputation management. Inf. Sci. **367–368**, 159–175 (2016b)
10. Eyal, I., Sirer, E.G.: Majority is not enough: Bitcoin mining is vulnerable. In: International Conference on Financial Cryptography and Data Security, pp. 436–454. Springer, Berlin (2014)
11. Germann Molz, J.: Collaborative surveillance and technologies of trust: online reputation systems in the 'new' sharing economy. In: Jansson, A., Christensen, M. (eds.) Media, Surveillance and Identity: A Social Perspective, pp. 127–144. Peter Lang, New York (2014)
12. Hamari, J., Sjöklint, M., Ukkonen, A.: The sharing economy: why people participate in collaborative consumption. J. Assoc. Inf. Sci. Technol. (2015)
13. Hildebrand, T., Puri, M., Rocholl, J.: Adverse incentives in crowdfunding. Working paper. http://papers.ssrn.com/sol3/papers.cfm?abstract_id=1615483 (2014)
14. Johnson, B., Laszka, A., Grossklags, J., Vasek, M., Moore, T.: Game-theoretic analysis of DDoS attacks against Bitcoin mining pools. In: International Conference on Financial Cryptography and Data Security, pp. 72–86. Springer, Berlin (2014)
15. Kamvar, S.D., et al.: The EigenTrust algorithm for reputation management in P2P networks. In: Proceedings of the 12th International Conference on World Wide Web, pp. 640–651 (2003)
16. Kiayias, A., Koutsoupias, E., Kyropoulou, M., Tselekounis, Y.: Blockchain mining games. In: Proceedings of the 2016 ACM Conference on Economics and Computation, pp. 365–382. ACM, Boston (2016)
17. Liu, Q., Vasarhelyi, M.A.: Big questions in AIS research: measurement, information processing, data analysis, and reporting. J. Inf. Syst. **28**(1), 117 (2014). https://doi.org/10.2308/isys-10395
18. Mainelli, M., Milne, A.: The Impact and Potential of Blockchain on Securities Transaction Lifecycle. SWIFT Institute Working Paper (2016)
19. McKinsey & Co. Report : Global Banking Practice, Cutting Through the FinTech Noise: Markers of Success, Imperatives For Banks, Dec 2015, McKinsey & Company (2015)
20. Milne, A., Parboteeah, P.: The Business Models and Economics of Peer-to-Peer Lending. Chicago (2016)
21. Naehrig, M., Lauter, K., Vaikuntanathan, V.: Can homomorphic encryption be practical? In: Proceedings of the 3rd ACM Workshop on Cloud Computing Security Workshop, pp. 113–124. ACM, Boston (2011)
22. Narayanan, A.: Bitcoin and game theory: we're still scratching the surface, Blog post on Freedom To Tinker, 31 Mar 2015 by Arvind Narayanan. https://freedom-to-tinker.com/2015/03/31/bitcoin-and-game-theory-were-still-scratching-the-surface/. Accessed 12 Jan 2017

23. Petrasic, K., Bornfreund, M.: Beyond Bitcoin: the blockchain revolution in financial services, White and Case report. http://www.whitecase.com/sites/whitecase/files/files/download/publications/the-blockchain-thought-leadership.pdf. Accessed 2 Dec 2106
24. Preez, D.: Sharing economy = end of capitalism? Not in its current form, Blog post on diginomica, 17 Aug 2015. http://diginomica.com/2015/08/17/sharing-economy-end-of-capitalism-not-in-its-current-form/. Accessed 10 Jan 2017
25. Ricci, S., Domingo-Ferrer, J., Sánchez, D.: Privacy-preserving cloud-based statistical analyses on sensitive categorical data. In: Modeling Decisions for Artificial Intelligence-MDAI 2016, pp. 227–238. Springer, Berlin (2016)
26. Rifkin, J.: The Zero Marginal Cost society: the Internet of Things, the Collaborative Commons, and the Eclipse of Capitalism. Macmillan (2014)
27. Romer, D.: Advanced Macroeconomics. McGraw Hill, New York (2011)
28. Salaman, G., Storey, J.: A Better Way of Doing Business?: lessons from The John Lewis Partnership. Oxford University Press (2016)
29. Tolentino J.: How blockchain is transforming business models, blog post by Tolentino J. In: Reinventing the World of Banking. http://thenextweb.com/worldofbanking/2016/09/16/how-blockchain-is-transforming-business-models/. Accessed 3 Dec 2016
30. Turi, A.N., Domingo-Ferrer, J., Sánchez, D.: Filtering P2P loans based on co-utile reputation. In: 13th International Conference on Applied Computing-AC 2016, pp. 139–146. IADIS Press, Mannheim (2016)
31. Turi, A.N., Domingo-Ferrer, J., Sánchez, D., Osmani, D.: A co-utility approach to the mesh economy: the crowd-based business model. Rev. Manag. Sci. (to appear) (2016b). doi:10.1007/s11846-016-0192-1

Printed in the United States
By Bookmasters